KNOW YOUR EYES

KNOW YOUR EYES

IRA A. ABRAHAMSON, JR., MD
Assistant Clinical Professor of Ophthalmology
University of Cincinnati School of Medicine

MEDCOM PRESS
MEDCOM MEDICAL UPDATE SERIES

Medcom Production Staff

Publisher
Robert E. Fuisz, MD

Publisher
Richard C. Fuisz, MD

Editor in chief
David P. Lauler, MD

Medical director
Virginia Malfitan, MD

Executive editor
Robert B. Wilson

Managing editor
Ruth A. Rosenblum

Executive film producer
Alfred R. Kelman

Executive art director
Gary N. Olivo

Project editorial director
Betty Treiber

Medical editor
Patricia E. Franks

Production manager
Joan Milarsky

Art director
Edward Zytko

Graphic illustrator
Vantage Art, Inc.

Manager, photography and permissions
Helena Frost

Photo research
Cathie Furbush

Copy staff
Marion Lazos
Mary Bozeman Raines
Peter Parks
Selma Ramsey
Elena Rosti
Susan Rubin

Medcom Press is a division of MEDCOM, Inc.

MEDCOM®
LEARNING SYSTEMS
World leader in multimedia health education programs
2 Hammarskjöld Plaza, New York, N. Y. 10017
(212) 832-1400

Library of Congress Catalog Number 72-8814

To my dear father, Ira A. Abrahamson, Sr., MD—a great friend, teacher, and outstanding compassionate ophthalmologist with whom I've had the privilege and pleasure of practicing for the past 18 years.

PHOTO CREDITS

Figures 1.5; 3.4, 3.5, 3.7, 3.13, 3.14, 3.15, 3.16, 3.17, 3.18, 3.19; 4.1, 4.19, 4.35, 4.38, 4.41, 4.71, 4.77; 6.4, 6.5, 6.6, 6.7—Gordon Baer of Black Star. **2.1; 3.2, 3.3, 3.6**—American Optical Corporation. **4.73**—House of Vision. **3.11**—Ralph Sutton, Contact Lens Service. **4.65**—Dept. of Ophthalmology, University of Cincinnati. **6.3**—Jack Foster. **4.3, 4.4, 4.5**—Dermatologic Ophthalmology, Part I. Photos of private consultation with author—Bill Powers.

CONTENTS

PREFACE

SIGHT IS OUR MOST VITAL SENSE. We prize it, but we usually take it for granted. We fear losing it, but we really know little about preserving it. In fact we may know little about how we see. What is the structure of the eye? How does the eye function? What does an eye examination reveal? How do glasses and contact lenses correct faulty vision? What happens when certain parts of the eye are damaged by disease or injury? Can this damage be repaired? What are the myths and misconceptions about the eyes? What constitutes good eye care? What are the basic tenets of emergency care to the eye? Knowing the answers to these questions will broaden your vision.

1

STRUCTURE OF THE EYE

THE EYE AND ITS RELATED STRUCTURES constitute a complex and well-coordinated unit. Each intricately designed part of this unit participates in or protects the function of sight.

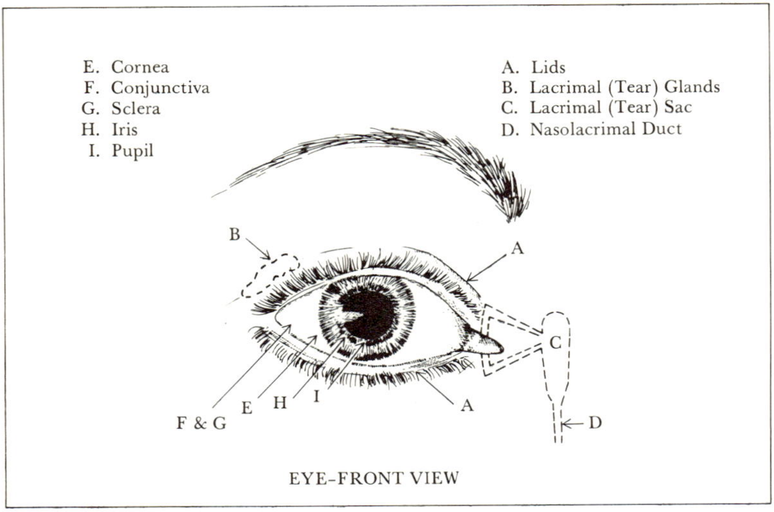

Fig 1.1

LIDS

The lids (1.1) open and close by reflex or voluntary action to distribute tear fluid, to shut out light, and to protect the eyes from foreign bodies and exposure. Blinking keeps the cornea, the transparent outer layer of the eyeball, moist and clear even when the eye is exposed to wind, dust, or dry air.

The outer surface of the lids is a layer of skin continuous with the skin of the forehead above and that of the cheeks below. This outer layer contains muscles that elevate and lower the lids, a firm tissue plate, or tarsus, that maintains their shape, and lashes that prevent perspiration or small foreign bodies from entering the eye and damaging the transparent sensitive surface of the cornea.

The inner lid surface is lined with a thin, transparent, cellophane-like mucous membrane called conjunctiva that also extends over the front of the eyeball to the cornea. The membrane lining the lids is palpebral conjunctiva; the membrane over the eyeball is bulbar conjunctiva. At the junction of these membranes is a fold known as the conjunctival fornix, or cul-de-sac. The conjunctiva has a rich supply of blood, which accounts for the appearance of the bloodshot eye after irritation. It also contains lubricating glands that permit the lids to move easily and the eye to rotate smoothly.

The outer junction of the upper and lower lids is the temporal canthus or angle; the inner junction near the nose is the nasal canthus. Just inside the nasal canthus is a small elevation of tissue, the caruncle. A skin fold at the inner angle, the epicanthal fold, is usually present in all children. Although a child with a prominent epicanthal fold may appear to have crossed eyes, the fold usually disappears by the time he is 4 to 6 years of age. The epicanthal fold is a normal characteristic in Oriental races.

ORBITS

The eyes are set in bony walled cavities or sockets called orbits that protect them from the injury of outside impact. The cone-shaped orbits are lined with a tough membrane, or periosteum, and contain a pad of fat, lying just behind the eyeball, which acts as a cushion as well as a support.

Also within the orbits are the muscles that move the eye, the many blood vessels that nourish it, and the nerves that control it. The optic nerve extends from the eye back through a small opening in the orbit, the optic foramen, into the brain. Both blood vessels and nerves pass through this as well as other similar openings or fissures.

The orbit also forms the wall to the adjacent sinuses. Sinuses are air spaces in the skull lined by the same kind of membrane as the nose. Canals connecting the sinuses to the nose allow secretions and tears to drain through the nose. The frontal sinus is located in the forehead; the maxillary, in the cheek; and the ethmoidal, alongside the bridge of the nose. Since the nerve pathways of the eye run close to these sinuses and the teeth, sinusitis or bad teeth can produce swelling of the lids or pain in the eye.

LACRIMAL SYSTEM

The lacrimal system is a highly efficient irrigation works. A steady flow of tears, or lacrimal fluid, keeps the eyes moist, protecting them from the drying effect of air while the lids are open. Lacrimal fluid also is capable of destroying infectious bacteria and neutralizing mild acids and alkalies.

Tears are produced in the lacrimal glands (1.1) which, like sprinklers, distribute the fluid over the surface of the eye. The tears then drain into the lacrimal duct system. First they enter a small hole, the punctum, in the lower lid near the nose. Then they pass through a tube, the canaliculus, and collect in the lacrimal sac in the inner corner by the nose. From this sac or pouch, they flow through the nasolacrimal duct into the nose, where they evaporate on the warm surface of the mucous membrane.

When the eyes are irritated by smoke or onions or when a person cries, tears are overproduced, and the result is a runny nose. When the nerve to the tear gland is paralyzed or when the gland is damaged by disease, tears are underproduced and the result is dry burning eyes; these symptoms are often relieved by artificial eye drop tears. When the tear drainage passage is blocked anywhere along its course due to infection, inflammation, or a birth defect, excess tearing will naturally result, and corrective surgery may be required.

EYE MUSCLES

The eyeballs are held in their orbits by a set of six muscles attached to the outside of each eyeball and anchored to the walls of the orbit. These muscles allow the eyes to move freely and harmoniously in all directions. Their location on the eyeball and their pulling functions are:

- Medial rectus (MR). Attached to nasal side of eyeball. Pulls eye toward nose.
- Lateral rectus (LR). Attached to temporal side of eyeball. Pulls eye out and away from nose.
- Superior rectus (SR). Attached to upper part of eyeball. Raises eye by pulling toward forehead or when looking toward temple; pulls eye toward nose and rotates eyeball inward in clockwise fashion when vision is directed toward nose.
- Inferior rectus (IR). Attached to lower part of eyeball near cheek. Lowers eye by pulling toward cheek or when looking toward temple; pulls eye toward nose and rotates eyeball outward in counterclockwise fashion when vision is directed toward nose.
- Superior oblique (SO). After passing through pulley-like structure (trochlea) on medial anterior orbital wall, at-

Fig 1.2

EYE MUSCLES

tached to outside and back of eye below SR. Lowers eye by pulling toward nose; rotates eye inward in clockwise fashion.

• Inferior oblique (IO). After passing from inferior orbital rim below IR, attached to outside and back of eye beneath LR. Raises eye by pulling toward nose; rotates eye outward in counterclockwise fashion.

The eyeballs respond to the pull of the eye muscles like a team of horses with reins on both sides of their heads (1.2). When the left reins are pulled, both horses turn left; when the right reins are pulled, both horses turn right. However, the movements of the six sets of muscles are more complex, more well-coordinated, and more subtle than those of the reins. For example, when you look to the left, the MR of the right eye and the LR of the left eye contract, as the LR of the right eye and the MR of the left eye relax. When you look at a very close object, the MR of both eyes contracts, as the LR of both eyes relaxes. Moreover, the "reins" of the eye can rotate the eyeballs as well as pull them sideways or up and down.

The nerve supply to the eye muscles comes from three of the body's 12 cranial nerves. The third cranial nerve (oculomotor) supplies all the muscles except the SO, which is supplied by the fourth cranial nerve (trochlear) and the LR, supplied by the sixth cranial nerve (abducens).

EYEBALL

The eyeball (1.3) is normally an oval body a little smaller than a ping-pong ball or abôut one inch in diameter. Myopic (nearsighted) eyes are generally somewhat longer, and hyperopic (farsighted) eyes are somewhat shorter. Absence of the eyeball from birth is anophthalmia. Exceptional smallness of the eye is microphthalmia, and unusual largeness is macrophthalmia.

The wall of the eyeball is composed of three layers (1.3), each serving a different purpose.

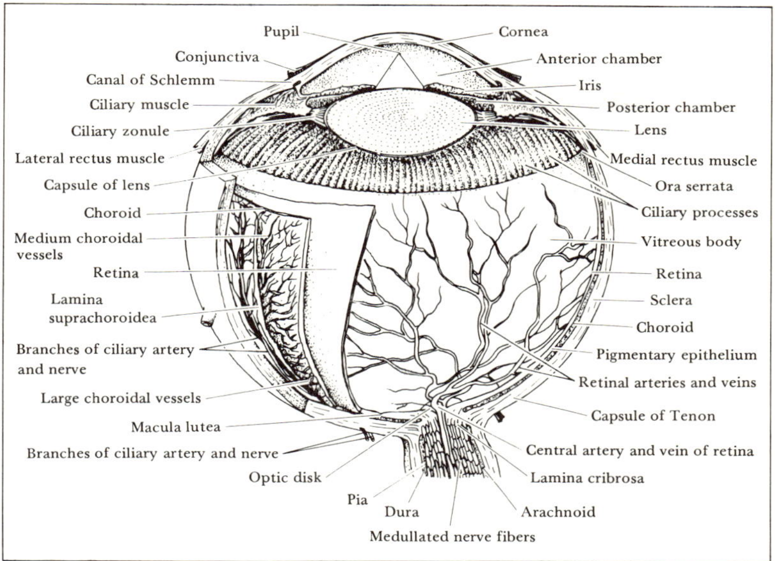

Fig 1.3

Sclera

The sclera, or the white of the eye, forms the outer layer. This tough opaque coat protects the contents of the eye and maintains the form of the globe. The external muscles are also attached to the sclera. In front of the sclera is a round opening. The transparent cornea covers this opening like the glass lens of a sealed-beam headlight.

Choroid

The choroid forms the middle layer. This thin dark-brown spongy coat consists mainly of blood vessels that supply the eye with nutrition and carry away waste products. The dark color of the choroid is produced, in part, by the many pigmented (colored) cells that form the background from which images are projected to the brain. In the front part of the eye the choroid joins the muscular structure, or the ciliary body. From this structure extends the iris or the colored part of the eye.

The iris, the ciliary body, and the choroid together make up the uveal tract. The outer layer of the ciliary body produces the watery fluid known as aqueous humor. The color of the iris depends on the amount of pigment. Albinos have no pigment, and the iris appears pink or red; blue or green eyes have little pigment; and brown eyes have a considerable amount. The black space in the center of the iris is the pupil.

The iris itself contains two circular muscles: the sphincter and the dilator. When the sphincter muscle constricts, the pupil becomes smaller; when the dilator muscles contracts, the pupil becomes larger. When the dilator muscle contracts, the constrictor muscle relaxes, and vice versa. By dilating or constricting, the pupil controls the amount of light entering the eye, as the shutter does in the camera. On bright sunny days, or in bright light, the pupil may constrict to pinpoint size. On dull days, or at dusk or nighttime, it dilates to provide more light for adequate vision. If you look into a mirror while flashing a light into one of your eyes, the pupil will constrict as the light strikes it.

Retina

The retina forms the innermost layer of the eyeball. This delicate transparent coat contains nerve cells and fibers. It is photosensitive (sensitive to both light and darkness) and obtains its blood supply from the choroid. A great abundance of nerve cells, or cones, is located in the central seeing part of the retina, the macula lutea. Vision is sharpest in the center of the macula, the fovea centralis (1.3 and 1.4). The fovea functions best in the daytime when the pupils are constricted. (Similarly the focus of a camera is sharper when the lens aperture is reduced.) The cones are responsible for sharp central vision, color, and form.

Other nerve cells called rods are present in the outer

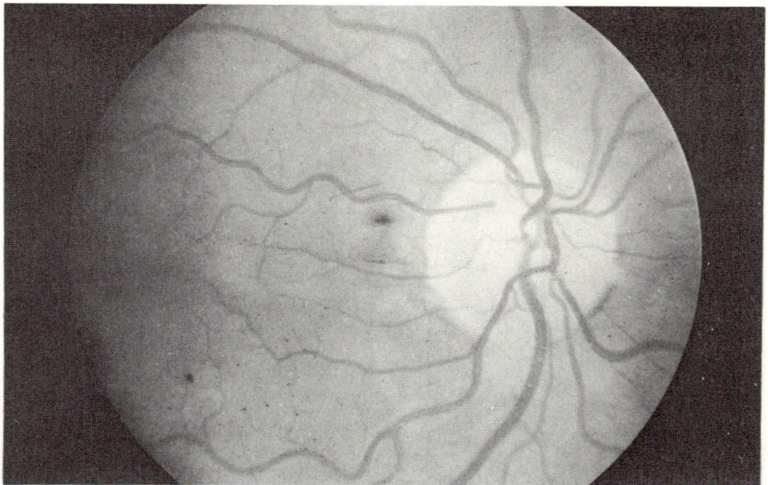

Fig 1.4

layers of the retina, and these function best at night when the pupil is dilated. Rods are responsible for peripheral (side) vision.

Images from the outside world are perceived by both rods and cones, and these images are sent through the retinal nerve fibers to the optic nerve, which extends like a stem from the back of the eye to the brain. The retina and its blood vessels may be observed with an instrument called an ophthalmoscope (1.5). The back part of the eye seen with this instrument is the fundus (1.4).

Lens

Within its three-layered walls the eyeball contains the transparent crystalline lens. The function of this lens is to focus rays of light entering the eye onto the retina, the innermost coat (1.3). The size and shape of the lens vary; the lens may become as small and round as a pea or as large and disklike as a small aspirin tablet. Enclosed in a fine thin capsule that prevents the watery aqueous humor from entering its

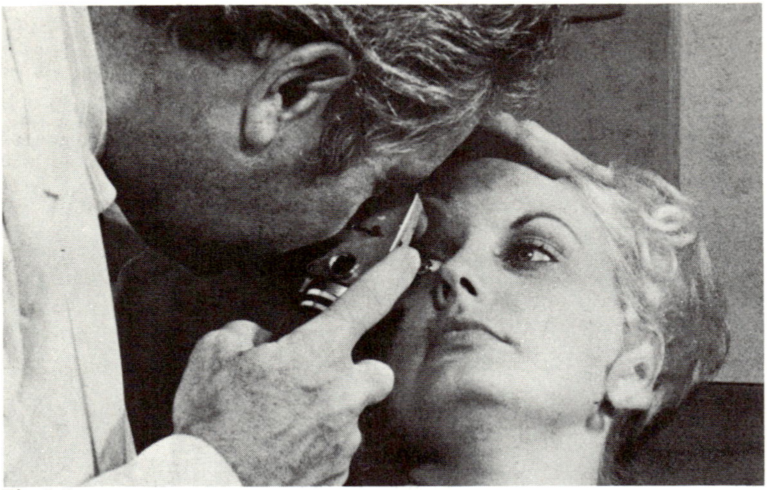

Fig 1.5

substance, the lens is held in position behind the iris by thin cellophane-like elastic tissues called suspensory ligaments or zonules, which extend from the lens to the ciliary body.

Vitreous

Behind the lens lies the vitreous, a transparent jelly-like substance that fills the entire back portion of the eye, or four fifths of the whole eyeball. In young people the vitreous is usually rather solid, having the consistency of a gelatin dessert in the summer; with age or with certain diseases it becomes more liquid. If lost through surgery or injury, the vitreous will not regenerate.

Anterior Chamber

In front of the lens the space between the cornea (in front) and the lens and iris (behind) is the anterior chamber. This chamber contains the aqueous humor. This fluid, if lost through surgery or injury, will reappear. The angle formed by the iris root and the scleral-corneal junction contains a circular channel known as Schlemm's canal, which serves as a drainage mechanism for the eye. The filtering of the fluid within the eye through these fine channels or openings helps to regulate tension or maintain pressure within the eyeball.

The Eye, the Camera, and the Watch

Some of these eye structures may be compared to certain parts of a box camera or a watch (1.6).

• The cornea is like the transparent cap over the lens of the camera or the crystal of the watch, while the sclera is like the outer covering of the camera or the outer body of the watch.

• The anterior chamber is similar to the space between the crystal and the face of the watch.

• The angle of the eye corresponds to the junction between the crystal and face of the watch.

Fig 1.6

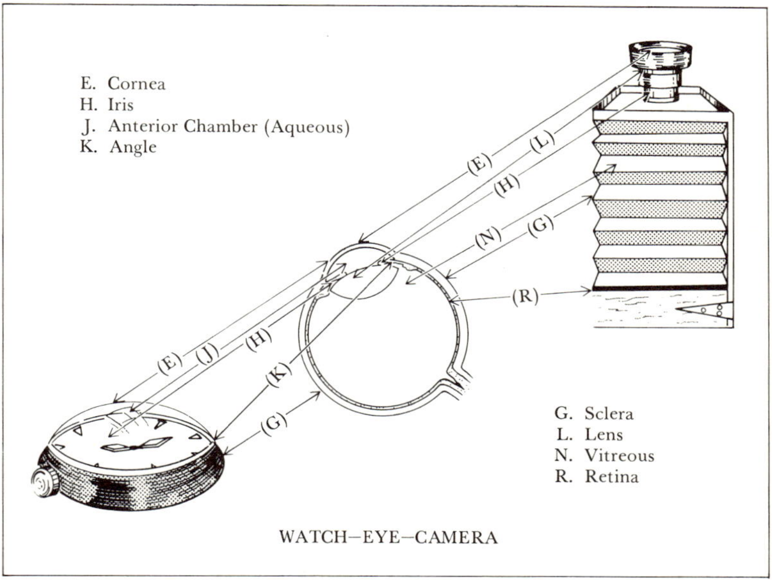

- The iris and pupil are like the shutter of the camera.
- The lens of the eye, located in the black pupillary space, resembles somewhat the lens of the camera. However, the flexible lens of the eye can alter its shape by becoming either flat or thick.
- The vitreous humor, filling four fifths of the eyeball behind the lens, corresponds to the hollow space in the camera.
- The retina, or the neurosensitive seeing part of the eye, resembles the photographic plate or film of the camera.

2

FUNCTION OF THE EYE

THE EYE IS THE SENSORY ORGAN OF VISION, but the brain converts the images that the eye sees to informative sight.

HOW DO WE SEE?

When light enters each eye in a parallel plane, it passes first through the cornea, the transparent covering of the eye, and then through the lens, which bends the light rays. After continuing through the jelly-like vitreous, the rays focus on the back of the eye, the highly sensitive retina, at the fovea centralis or center of acute detailed vision.

In the pigmented or colored layer of the retina photochemical and photoelectrical responses convert the light rays to electrical impulses that stimulate the rod and cone cells. The picture, which is inverted when the retina receives it, is then transmitted through the retinal nerve fibers to the optic nerve of each eye.

As the optic nerves leave the eyes, they pass through the orbits to the brain. Here the right and left optic nerves join each other in the optic chiasm. Some fibers from the left eye go to the right side of the brain via the optic tracts. Nerve fibers from the nasal (inner) part of the retina of the left eye and the temporal (outer) part of the retina of the right eye go through the right optic tract, then on through the optic pathways or radiations to the right occipital area in the back of the brain. The pattern is reversed for fibers reaching the left occipital lobe of the brain.

In the occipital lobe the image is turned right side up, and the actual interpretation of conscious sight begins. The brain interprets the image by discerning both the light and the form of the stimulating agent, and it records the image.

This interpolation of sight is similar to the phenomena of television. Objects are recorded first by the television camera (light rays focused on the retina). The electrical patterns of television (electrical responses in the eye) are then transmitted through bundles of conductors and cables (retinal nerve fibers and optic nerve) until they appear on the television screen (occipital lobe of the brain) where they are interpreted and seen as a clear picture.

DAY AND NIGHT VISION

The iris-pupil diaphragm, like the shutter in a camera, regulates the amount of light entering the eye by constricting or narrowing, or dilating or widening. In bright light the pupil becomes small; in darkness or in poor light the pupil enlarges, permitting more light to enter the eye. This pupillary action is connected intimately to the photosensitive seeing elements of the eye, the rod and cone cells in the retina.

The cone cells are located primarily in the fovea centralis, which is situated in the center of the macula lutea, and they are responsible for acute central vision. Cones function best in the daylight with a small pupil, and this sight phenomenon is called day (photopic) vision. Disease in this central area may cause poor central sight, but not blindness. These people often see worse in the daytime with constricted pupils, and better in the night or in the dark with dilated pupils or by shifting their vision away from the object to be seen.

The rod cells are situated mainly in the periphery of the retina, and they function best in the darkness with a widely dilated pupil. Rods are responsible for night (scotopic) vision as well as for side (peripheral) vision. Disease of the rod cells may cause night blindness, difficulty in seeing in the dark or while attending movies, or constriction of peripheral vision. The temporary sensation of pain in the eyes when coming out of a movie into bright light is caused by the sudden constriction of the dilated pupils.

COLOR VISION

The phenomenon of color vision takes place in the cone layer of the macula lutea in the retina. The three primary colors in the spectrum—red, yellow, and blue—combine to form the basic formula for all other colors. A person who can perceive all colors accurately is called a trichromat. If the macula lutea is damaged by disease or injury, color perception may be impaired.

COLOR BLINDNESS

Color blindness is a defect in the perception of the normal color pattern. A color-blind person has a partial or total inability to distinguish between such component colors as red and green or violet and blue. To test color blindness Ishihara and Hardy, Rand, and Rittler developed a book of pseudoisochromatic color plates. The plates represent numbers and figures made up of carefully selected color dots blended with dots of contrasting confusing shades (2.1). Since the color-blind person is confused by these colors and shades, he will not see a number at all or will see a different number. The type of color blindness is determined by observing which colors the person cannot identify on a series of cards. In another test the person is asked to distinguish various shades of thread.

Color blindness is usually hereditary. Women serve as carriers of the defect; they transmit this condition to their children, but are not usually color blind themselves. Men

Fig **2.1**

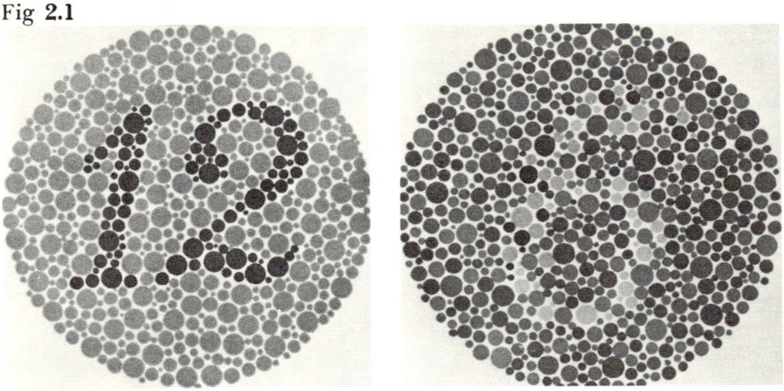

actually exhibit the defect. There are no drugs or exercises to cure color blindness; the only therapy is reeducation. For example, a person may be taught to recognize the position, rather than the color, of the red or green light in a traffic signal. However, even though reeducation enables a color-blind person to distinguish colors, his basic pattern of color discrimination cannot be changed.

3

EXAMINATIONS OF THE EYE

EXAMINATIONS OF THE EYE are carried out for different purposes by different members of the eye care team.

THE EYE CARE TEAM

Ophthalmologist (derived from the Greek *ophthalmos,* eye, and *logos,* discourse). This doctor of medicine (M.D.) has the same training and background as a family physician plus three or more years of specialized postgraduate work in ophthalmology. He is usually qualified for or certified by the American Board of Ophthalmology. As an eye physician, he is fully trained to treat eye diseases and perform eye surgery, as well as to carry out examinations for glasses, fundus examinations, visual field tests, and measurements of intraocular tension.

Oculist (derived from the Latin *oculus,* eye). This term was adopted by the early American glass fitters and opticians. Today it is an obsolete term which is synonymous with ophthalmologist.

Optometrist (derived from the Greek *optikos,* of the sight, and *metron,* measure). This optometry doctor (O.D.) has special training above the college level to enable him to examine eyes for glasses, to prescribe and fit glasses, and to perform visual training for development of ocular coordination and increased visual efficiency. He cannot use medication in the eyes, treat diseases, or perform surgery. Because he is not a medical doctor, he refers patients with eye disorders to the ophthalmologist.

Optician (derived from the Latin *opticus,* optics). This technician grinds lenses and fits glasses according to the ophthalmologist's or optometrist's prescription. He bears the same relationship to the ophthalmologist that the bracemaker does to the bone surgeon or the pharmacist does to the physician.

EXAMINATION FOR GLASSES

When should a child or an adult have an examination for glasses? When he often has a headache or tired eyes? When he can't quite see the spelling words on the blackboard in his classroom or the STOP sign at the end of his busy street? When his vision seems normal, and he has no complaints?

If children were examined routinely at age four or five or before they entered kindergarten, many eye defects would be discovered, diagnosed, and corrected before complaints became annoying or before changes became irreversible. Often a young child does not realize that his vision is impaired; he may not complain even if he has a headache or tired eyes. For these reasons, many communities have adopted preschool screening programs utilizing the services of trained lay people to rule out the presence of decreased vision (amblyopia), lazy eyes, and squint or strabismus.

Even though children undergo routine screening after they enter school, these examinations are not complete. Although the child's visual acuity is tested, he may still need glasses or have an eye disease. A child should have his vision tested every year, but it is best to have his eye doctor give him a complete examination. Adults should be examined completely every one to three years.

Testing Visual Acuity

Visual acuity is tested by having a person stand or sit at a given distance, usually 20 feet, from a specially designed chart, usually the Snellen chart (3.1, 3.2). If the distance is ten feet, the chart may be placed over the person's head, while he looks at a mirror ten feet away. In actuality the distance will be 20 feet. A projector type of chart produces the same effect with distances less than 20 feet. The chart (3.2) has standardized words, pictures, numbers, or "Es," varying in size from large to small. Each character subtends an angle of 5′ (one twelfth of a degree) to the eye, as the observer would see it from various distances (3.1). The smallest character that he can read with each eye separately is recorded. The uppermost letter or figure is designed to be read by

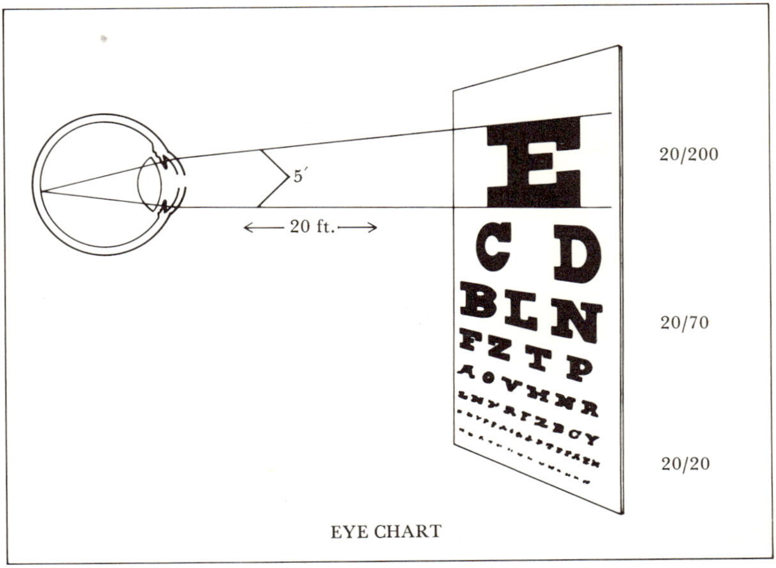

Fig 3.1

the normal eye at 200 feet; the rows following that should be read at 100, 70, 50, 40, 30, 20, 15, and 10 feet.

For distance vision 20/20 is average or normal vision. This means that a person can stand at 20 feet and see what he should normally see at 20 feet. (The numerator of the fraction represents the person's distance from the chart; the denominator represents the distance that corresponds to the letter size he can see.) Thus 20/40 vision means that a person must come up to 20 feet to see what he should see at 40 feet, or that he is seeing the "40-foot" letter size at 20 feet. On the other hand, 20/15 vision is better than average vision. In clinics using meters rather than feet to measure distance, 6/6 is equal to 20/20 and 6/12, to 20/40.

Vision is recorded as:

O.D. or RE (oculus dexter or right eye) or R = 20/15
O.S. or LE (oculus sinister or left eye) or L = 20/15

If a person cannot see the big "E" at 20 feet, he walks toward the chart until he sees it. If he sees the "E" at ten feet,

TEST *YOUR* VISION

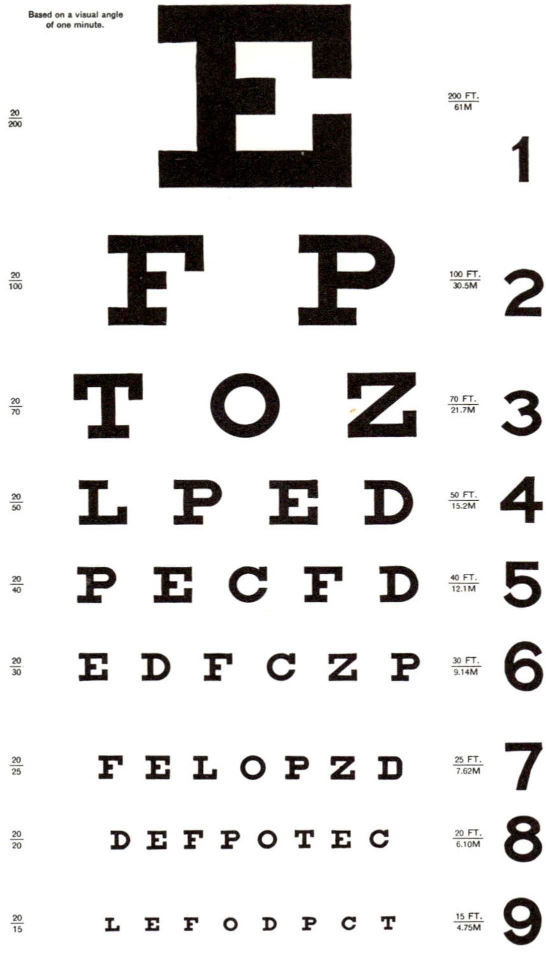

Fig 3.2 Snellen Chart

Do you need glasses? Take this quick test to see just how sharp your eyesight is.

Place the book upright on a table or chair, or have someone hold it for you. Stand 5 feet 10 inches away from the chart. *Test each eye separately* by covering one eye and then the other.

If you can read line eight (the next to the last line), your vision is normal, or 20/20. If you can read the bottom line,

STANDARD TEST TYPES

Arranged by DAVID W. WELLS, M.D.

4 Point — 1 — Small Bible

The only accurate way to measure sight is by means of letters or carefully graded characters, viewed at a distance of twenty feet. Quite a variety of such cards have been devised by eminent doctors. Near test types are used to determine the patient's ability to see to read at the proper distance. Type as small as this is not in general use for books or papers, except where lack of space necessitates it. The terms used by

5½ Point — 2 — Newspaper

The terms used by printers to designate the different sizes are adopted, and familiar examples given of the customary use of each. Spacing between types increases legibility, therefore, the "leading" has been made to conform with standard typography. Newspapers are usually printed in 5½ and 7 point. Most magazines use larger type.

7 Point — 3 — Newspaper

This paragraph and the one before it are set in the style and sizes of type that are frequently used in newspapers. The 5½ point size is used in the sports and market pages, and the 7 point in news columns. Newspaper types are smaller than book types and they are used here to make this test.

10 Point — 4 — Text Books

Books should be printed on dull finished paper. It is very unfortunate the publishers of text-books for schools and colleges so frequently ignore this fact, in order to get good impressions of illustrations for which a glossy finish is needed.

11 Point — 5 — Books

For prolonged use of the eyes the type should be several sizes larger than the smallest which can be read. The ordinary book is printed in 10 point or 11 point, but in order to read this comfortably one should be able to read 5½ point.

12 Point — 6 — Books

In order to get a proper illumination one should sit with his back to the light. Objects are seen by the light which goes from the object to the eye, not from the eye to the object. This precaution is quite commonly neglected.

18 Point — 7 — Children's Books

Children should be allowed to use only such books as are printed in large, clear type, and excessive reading forbidden.

American Optical
COMPANY

9012 — No. 11966 — PRINTED IN U.S.A.

Fig 3.3 Jaeger Card

your vision is 20/15—or better than normal. However, if you can't read the lines above these, you should see your eye doctor. It may mean you need glasses.

To test your near vision, hold this book 16 inches away. You should be able to read all type, especially the top lines, which are the type size of a small Bible. This Jaeger card is reproduced at its normal size. If you have any difficulty reading the top three lines, see your eye doctor.

his vision is recorded as 10/200. If he cannot see the "E" at one foot, then the distance is noted at which he can count fingers (CF). If he cannot count his fingers, then the distance for hand movement (HM) is recorded. If he cannot see hand movements, then a light is flashed into his eyes from different directions to determine if he can see the light or tell the direction from which it is coming. This test is called light perception and projection (LP & P). If he can see the light, but cannot tell the direction from which it is coming, or the field of vision being stimulated by the light, vision is recorded as light perception (LP). Finally, if a person is totally blind, with no light perception, his vision is recorded as No LP, and his condition is called amaurosis.

To test near vision, a Jaeger type card (3.3) featuring paragraphs of type in letters of graduated size is used. A person is asked to determine the smallest printed line that he can read. Both the line and the distance that the card is held from the person are recorded. If he can read the line of smallest print at 14 inches, his vision is recorded Jaegeri or Ji, or P_1 or Point 1, which is normal (14/14 vision in inches for near vision corresponds to 20/20 vision in feet for distance). Persons who are presbyopic (unable to focus on nearby objects) are usually past 40 years of age. They may have Ji recorded at 20 inches or more, or possibly Jii, or Jiii, at a given distance.

Searching For Refractive Errors

REFRACTION

Although the term refraction normally means bending of light rays, here it will describe the examination of the eyes to determine the need for glasses or lenses to correct faulty vision. In performing this examination the doctor flashes light into the patient's eyes from an instrument called a retinoscope (3.4). Light reflexes projected back from the retina of the patient's eye are observed by the doctor through a small hole in

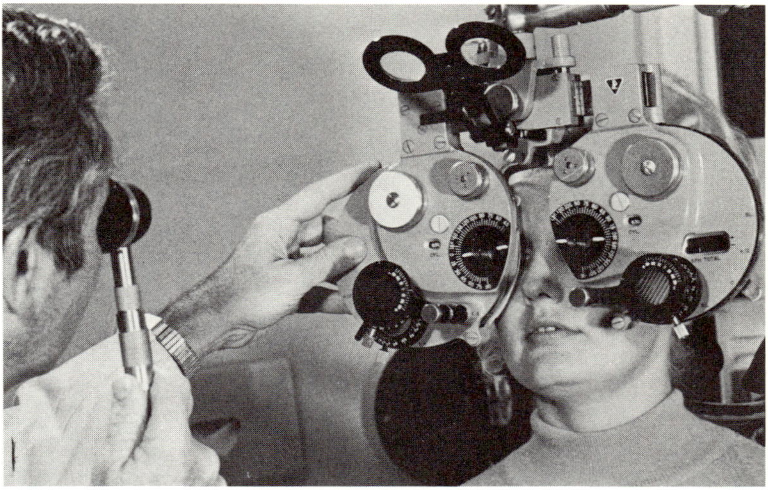

Fig 3.4

the reflecting mirror of the retinoscope. He then inserts temporary lenses into a phoropter (3.4), which is a battery of lenses set in front of the patient, or he places a trial frame and separate lenses (3.5) before the patient's eye.

The purpose of refraction is to neutralize the movement

Fig 3.5

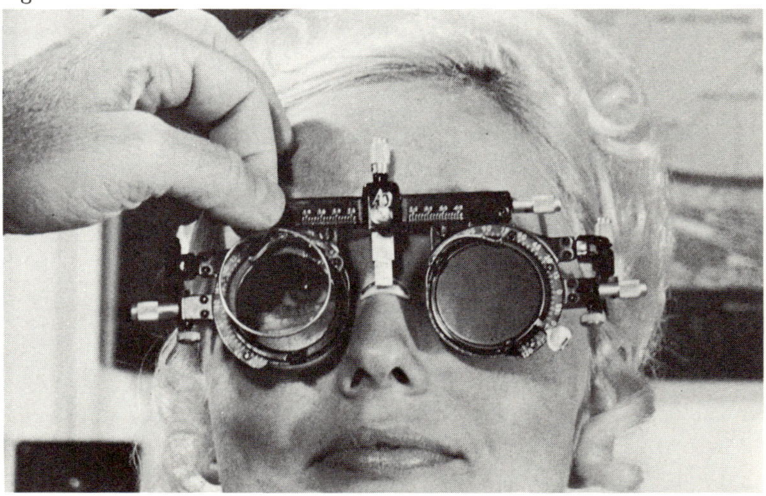

of the emerging rays of light from the patient's eye so that the doctor can determine the prescription for glasses or the lens correction required to place a clear image on the center of vision in the retina. With objective refraction, the doctor can accurately assess the patient's need for glasses even without the patient's help. Thus, an infant, deaf mute, or person with a language difficulty can be tested.

With subjective refraction the doctor inserts lenses before the patient's eye and asks whether the lenses improve his vision or make it worse. This process is continued until the best vision has finally been obtained. When the critical point is reached, the patient's decision may be difficult. If dilating cycloplegic drops are used to put the eye at rest by paralyzing accommodation (the ability to focus on close objects), the technique is called static refraction. If cycloplegic drops are not used, it is called manifest refraction. Since cycloplegic drops may rarely precipitate a sudden rise in intraocular pressure or cause drug poisoning, their use is legally restricted to medical doctors who are trained to recognize and treat any complications.

A new technique, automotive refraction, may revolutionize current refracting practices. With this device a light is flashed into the patient's eye, and a computer calculates the refractive error and records it on a graph in a matter of seconds.

Regardless of the type of refraction used, a special test to determine astigmatism (a refractive error due to changes in the corneal or lens curvature) may be given. The patient looks at a spoke or a wheel of equally black lines called an astigmatic dial (3.6). If some lines appear blacker than others, cylindrical lenses used to correct astigmatic error are inserted before the patient's eye until all lines appear equally black.

EYE DROPS

Two types of eye drops may be used for refraction: mydriatic drops or cycloplegic drops. When mydriatic drops

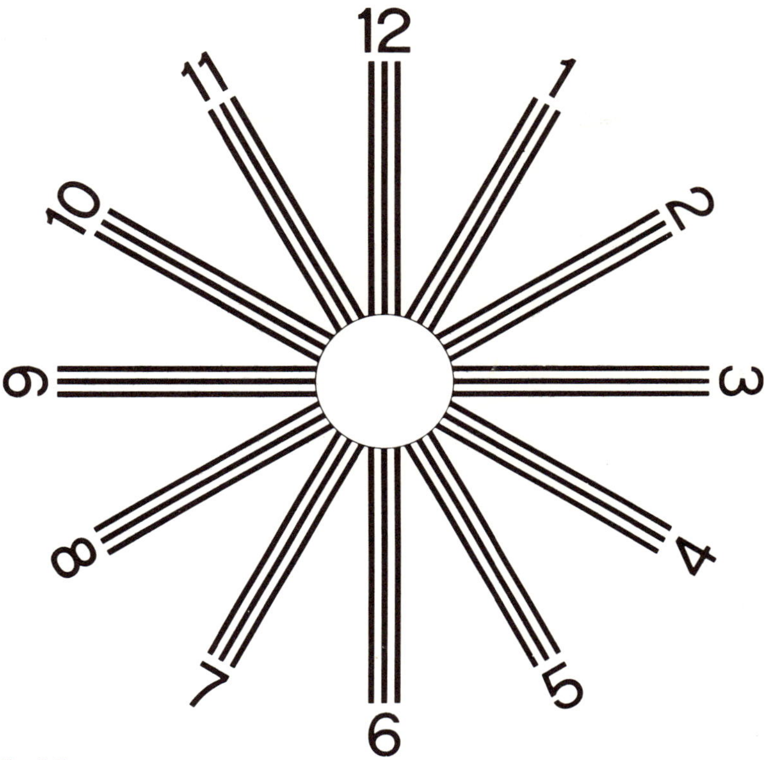

Fig 3.6

are instilled into the eye (3.7), the pupil becomes dilated or enlarged. When cycloplegic drops are instilled, the pupil not only becomes dilated, but the muscular structure called the ciliary body becomes temporarily paralyzed and the eye is put at rest. In a state of cycloplegia the eye temporarily loses its ability to focus on close objects, and hidden refractive errors may be revealed. This relaxation of accommodation also prevents the doctor from overcorrecting the vision of the myopic or nearsighted patient and from undercorrecting that of the hyperopic or farsighted person.

After receiving the drops, a person may have difficulty with his near vision or close work. If he is farsighted, his distance vision without the aid of corrective lenses will also be

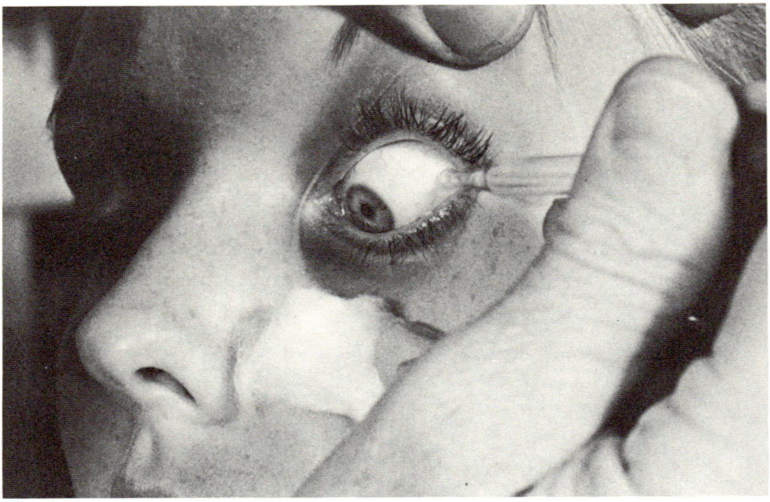

Fig 3.7

foggy. However, eye drops *do not blind* a person. Patients often ask, "Can I drive home after getting the drops?" The answer is: "Yes, if you were able to see well enough to drive to the office." Although distant vision is not affected greatly by the drops, a farsighted person whose near and distant vision become blurred while he is "under the drops" will experience more difficulty than the normal or nearsighted patient.

Depending on the type of drops used, the pupil may remain dilated for from six hours to 12 days. Mydriatic drops have a short duration of action (six to eight hours); cycloplegic drops have a longer lasting effect (24 hours to 12 days).

ACCOMMODATION

With accommodation, the crystalline lens within the eye changes its shape, permitting near or close objects to be seen clearly by focusing rays of light on the retina. The lens is held in position by zonular fibers or suspensory ligaments that run from the lens to the muscular ciliary body. As the ciliary body contracts, the zonular fibers relax, and the elastic lens becomes thicker and more globular in shape. This increases the power or strength of the lens. Through the mechanism of accommodation the eye can quickly bring near objects clearly into focus and then, in less than a second, it can focus on distant objects. As the ciliary body relaxes when the eyes look off into the distance or when the eye is put at rest during cycloplegia, the zonular fibers contract or tighten. This tension causes the lens to become flattened or less globular in shape, as the periphery of the lens is pulled towards the ciliary body; this is actually the normal position of the relaxed lens. If no refractive error is present, objects over 20 feet away can be seen clearly.

Accommodation is most powerful during infancy and youth. In fact the strong accommodative mechanism of children and young people may enable them to overcome a considerable refractive error and to read the 20/20 line when their vision is tested. However, as a person approaches age 40, the lens fibers lose their elasticity and accommodation becomes less effective. Since the focusing power of the lens becomes limited in this condition, known as presbyopia, glasses must be worn for clear close vision. Those with hyperopia are also limited in their ability to accommodate to see close objects, while those with myopia are limited in their ability to accommodate to focus on distant objects.

Correcting Refractive Errors

The normal or emmetropic eye measures one inch front to back. Rays of light focus sharply on the retina, and vision is 20/20. No glasses are needed, as both distant and near vision are excellent. The ametropic eye has a refractive error, and corrective lenses are required to improve or restore sight to normal.

To restore vision to normal, spherical or cylindrical lenses, or a combination of the two, are prescribed at the discretion of an experienced and skilled refractionist. Spherical lenses correct simple near- or farsightedness; cylindrical lenses correct astigmatism, which may also be near- or farsighted.

The power of the lens is graded and recorded in units called diopters. A lens that brings rays of light to a focus at a distance of one meter away is called a one diopter lens (1 D); a lens that focuses rays at a half meter is called a two diopter lens (2 D). Due to the inverse ratio the stronger the lens, the shorter the focal length. This relationship can be demonstrated by holding a plus convex lens above a leaf and bringing the sunlight to a focus point on the leaf, causing it to burn. The power of a lens may be as low as 0.12 D (a weak correction and small refractive error) or as high as 20.00 D or more (a strong correction and large refractive error). Plus (+) denotes farsightedness and minus (—), nearsightedness. If a patient's test results vary, he is asked to return for a second trial. Here is a typical lens prescription:

RE = +2.00 sph = +1.00 cyl ax 90
LE = —2.00 sph = —2.00 cyl ax 170

This example represents farsighted astigmatism in the right eye and nearsighted astigmatism in the left eye. The abbreviation "sph" (or "S") means a spherical lens; the equal sign denotes "combined with;" "cyl" (or "C") means a cylindrical lens; "ax" (or "X") is the axis of the cylinder; and 90 and 170 indicate the direction of the cylinder axis in correcting the astigmatism. If bifocals are prescribed for close work, an annotation like this appears at the bottom of the prescription:

Add +2.00 sph (or "S")

Depending on the added strength of the correction, this annotation will vary.

Corrective lenses have only one function: to focus rays of light on the back of the eye. Disease in the refractive media (the cornea or lens) may interfere with sharp vision by obstructing the passage of light rays to the retina, or the retina itself may be damaged by disease. Thus glasses can improve sight only if no disease is present in the eye and only if poor vision is a result of an eye defect requiring corrective lenses to restore normal sight.

Corrective lenses can offset these types of refractive errors: myopia, hyperopia, astigmatism, and presbyopia (3.8). Some of these defects are inherited in a dominant mode: if the parents are myopic, the child may well be myopic. However, other errors are recessive; the child will not necessarily have the same defect as his parents.

MYOPIA

The myopic eye is longer than one inch from front to back. Since rays of light focus in front of the retina (3.8), the image is blurred. Myopic persons are called nearsighted because they see objects close at hand clearly, while objects far away appear blurred or hazy. Myopia may be stationary or slowly progressive. In progressive myopia the eyes grow longer

Fig 3.8

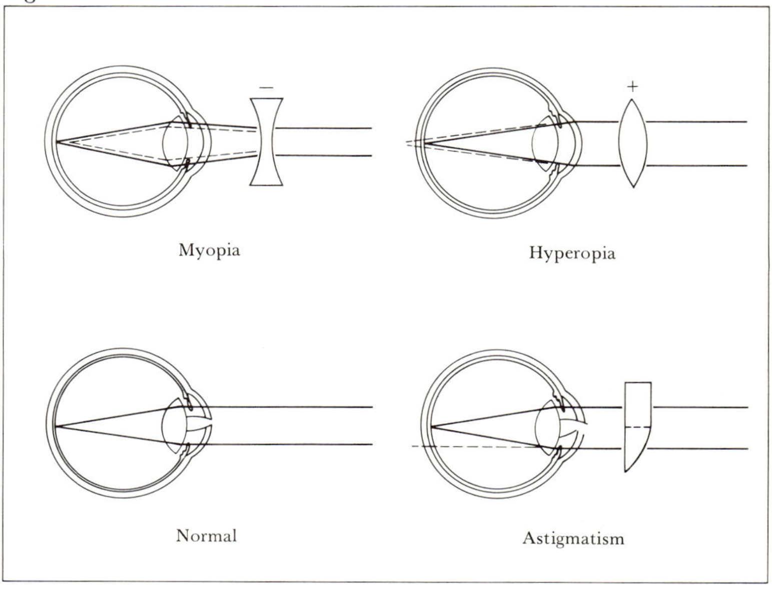

and larger, and the strength of lenses must be increased each year.

Myopia is corrected by a concave or minus lens (eg, —2.00 S) that lengthens the rays of light, bringing them to focus and fall on the retina (3.8). A minus lens can be identified by holding the lens in front of your eyes and looking through it at a straight line or object. If you move the lens from side to side, the object will move in the same direction.

Since persons with a high degree of myopia (over 6.00 D) have long eyeballs with the retina on stretch, they are predisposed to a detachment of the retina. Such persons should avoid blows to the eye, which may be incurred in boxing or high diving. When nearsighted persons reach the age of 40, however, they do have one advantage over their normal or farsighted contemporaries: they are still able to read or do close work without glasses.

When a child sitting at the back of a school room says that he cannot see the blackboard clearly, or when an adult cannot see a street sign at a distance, he has symptoms of myopia. If he squints his eye by narrowing his lids to a slit or if he looks through a tiny hole made by curling his thumb and forefinger together, he can see distant objects more sharply. Myopic persons may have larger and more prominent eyes with more dilated pupils than normal or farsighted persons. Since neither use nor disuse of the eyes by a nearsighted person alters the progress of myopia, the patient should be permitted to use his eyes as much as he wishes.

HYPEROPIA (HYPERMETROPIA)

The hyperopic eye is shorter than one inch from front to back. Since light rays focus behind the retina (3.8), the image is blurred. Hyperopia may also occur if the lens is removed at cataract extraction or if it is dislocated into the vitreous by an injury. A hyperopic person whose lens is intact is called farsighted because he can see things far away sharply, while objects close at hand appear blurred.

Hyperopia is corrected by a plus or convex lens (eg, +2.00 S) that focuses and shortens the rays of light, enabling them to fall upon the retina (3.8). Absence of the lens requires a strong plus lens to bring rays of light to focus on the retina. A plus lens can be identified by holding the lens in front of your eye and looking through it at a straight line or an object. If you move the lens from side to side, the object will move in the opposite direction.

Headaches, irritability, eyestrain, and fatigue after close work or use of the eyes are frequent complaints of farsighted persons. In children a high degree of uncorrected hyperopia may produce crossed eyes as well as poor vision. As the child gets older, hyperopia may diminish with growth, and less lens correction may be required.

ASTIGMATISM

In the astigmatic eye a defect in the curvature of the cornea or lens causes light rays to focus improperly on the retina (3.8). Regular astigmatism is an irregularity in one meridian of the circumference of the cornea; irregular astigmatism is an unevenness in different meridians. Astigmatism can be farsighted, nearsighted, or a combination of both called mixed astigmatism. It may result from scars to the cornea as well as from developmental abnormalities.

Astigmatism is corrected by cylindrical lenses (3.8) that may be plus (eg, +2.00 C X 90), minus, or mixed (eg, +1.00 S = −2.00 C X 90). A cylindrical lens can be detected by holding the lens in front of your eyes and looking through it at a straight line. If you rotate or rock the lens from side to side, the straight line will appear crooked. If the line moves with the lens, it is a minus lens; if it moves in the opposite direction, it is a plus lens. Astigmatic charts which have circles of radiating lines help to determine the degree and meridian of the astigmatism.

Eye fatigue, headaches, irritability, difficulty in focusing near or distant objects, and poor vision are symptoms associated with astigmatism. A clear image may be seen in one meridian and a blurred image in the other meridian; circles may become ovals; and lights may appear to have tails. In an attempt to improve his vision the patient may tilt his head to one side, or he may squint his astigmatic eye to cut out rays of the unfocused meridian.

PRESBYOPIA

The presbyopic eye cannot focus on close objects because the lens has become less elastic and has lost the ability to change its shape or increase its power. This weakening of the accommodative mechanism, which usually becomes most evident after age 40, progresses to age 70.

Presbyopia is corrected by the use of plus convex lenses (eg, +2.00 S) that are like magnifying glasses. Various forms of bifocals or reading glasses can be prescribed.

The presbyopic person finds himself holding his reading material further away than normal (3.9). He complains: "My arms are not long enough to see the paper," or "I cannot see the numbers in the telephone book clearly," or "The print is too small." Blurring of close work, fatigue after using the eyes, and headaches over the eyes after close work are other symptoms. Presbyopia produces visual complaints sooner in farsighted than in nearsighted persons. By removing his glasses to see close objects, the nearsighted person may temporarily escape the condition.

Fig 3.9

PRESBYOPIA

GLASSES

Among the types of glasses used to correct refractive errors are single-vision glasses, bifocals, and trifocals (3.10).

Single-vision glasses can correct any of the previously mentioned refractive errors, but they do *not* change the shape of the eyeball. Although the frame may be of any style, it should be fitted and adjusted properly and the lenses should be centered correctly. When the myopic patient wears glasses for the first time, he will think that the floor is far away from him. The hyperopic or presbyopic patient will feel that everything is too close. In a short time these symptoms will disappear. Although the presbyopic patient with reading glasses may be bothered by having to take off these single-vision spectacles to see across the room, that is the price he must pay unless he wants bifocals.

Bifocals combine a distance correction and a reading glass in one lens. In the executive type bifocal the lower third

Fig 3.10

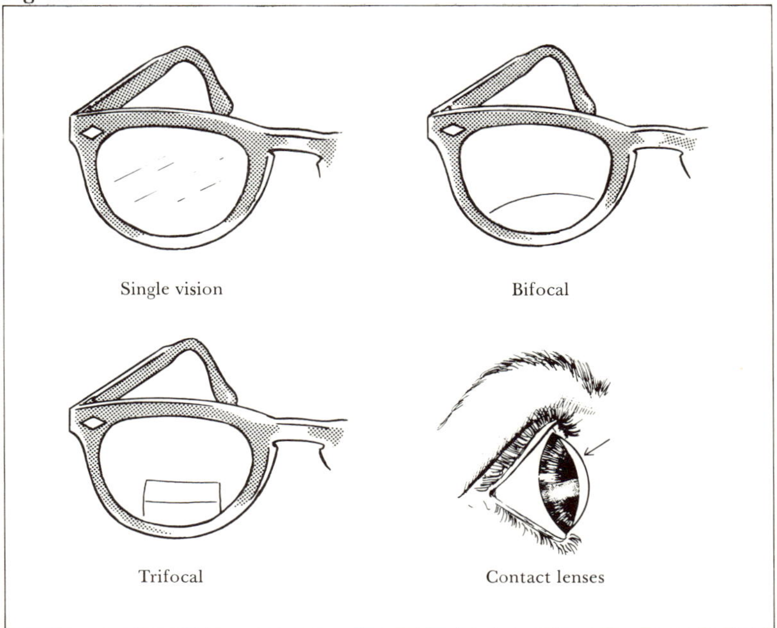

to half of the lens is for reading, and there is a visible line between the distance and reading glass. In the fused type bifocal the division between the two lenses is hardly visible. The Benjamin Franklin half type reading glass is becoming popular with those who have no distance correction, but who want the advantages of a lens to see close work and the freedom to use their naked eye for distance. For patients engaged in occupations requiring close work, a special working glass is available. The close correction lens is placed at the top instead of at the bottom of the glass, and it can be prescribed for various working distances ranging from 2 to 30 inches.

Difficulty in wearing bifocals may result from poorly fitting glasses or from the patient forgetting to reverse his previously developed reading habits. Normally when a person reads he lowers his head. With bifocals he raises his head, but he lowers his eyes to look down through the bifocal. The reverse is true when he looks off into the distance. Wearers of bifocals should exercise caution when going up and down steps and in crossing curbs. If bifocals are properly fitted, the patient will not become confused by the line separating the distance from the reading lens. However, wearing bifocals for the first time does require care and patience.

Trifocals provide three ranges of focus—distant, intermediate, and near—in one lens. They are advantageous for bankers, druggists, clerks, grocers, shoe salesmen, engineers, and secretaries, who are required to see objects at 15 inches, 22 inches, and in the distance. The new omnifocal lens provides the same ranges of focus as the bifocal or the trifocal. It is a single glass lens without any visible lines.

Proper fitting of glasses. The three types of forward glasses—single-vision glasses, bifocals and trifocals—should all be centered and fitted properly, with special emphasis on nose and ear adjustments, to insure maximum comfort. Some children need glasses, but do not want to wear them for cosmetic reasons or for fear that other children will make fun of them. For every such child, there are those who are found on examination to have no need for glasses, but want to wear them, because they think they will look more intelligent or glamorous. The doctor is happy when he gets "the right shoe on the right foot."

Safety and care of lenses. All lenses should be made of a shatterproof plastic or case-hardened material. Hopefully this strong recommendation will soon be a legal requirement. Nonbreakable safety lenses are prescribed not only for athletes or those in hazardous occupations, but also for children and one-eyed persons. Plastic lenses scratch more easily than those made of glass. Never lay glasses down with the lenses touching a table or other hard surface. Wipe lenses with a clean soft tissue or cloth, never with a coarse handkerchief or a rough towel.

Incorrect glasses. Wearing incorrect glasses will *not* produce any permanent eye damage or any organic disease. A wrong lens may cause headache, eyestrain, dizziness, discomfort, blurring of vision, or burning of the eyes. In response to these irritations a patient may rub his eyes and instigate an infection such as blepharitis or sties.

Constant wear of glasses. Glasses do not have to be worn all the time. Going without them will not damage the eye or aggravate the eye condition. However, glasses are prescribed to improve visual defects. If glasses are not worn, sight remains impaired, and the person may experience headache, eyestrain, dizziness, and other signs of poor vision.

Frequent changes of glasses. A young child may be farsighted before he reaches the age of six, but as his body grows he may eventually become nearsighted. His inherent pattern of growth determines both the shape of the eyeball and the correct lenses required for good sight. Since eye changes may be rapid during the growth period, children should have their eyes examined every six months to one year depending on the individual. Most nearsighted children are avid readers since their close vision is good while their uncorrected distance vision is hazy. Although teachers often think these children are exceptional students because they read so much, their grades may be poor because they are unable to see the blackboard clearly.

Frequent changes of glasses in adults may indicate the presence of cataracts, glaucoma, or diabetes. A thorough investigation by an ophthalmologist is recommended.

CONTACT LENSES

Contact lenses are glasses fitted to the cornea. Made of a plastic material, they are smaller and thinner than the fingernail of the little finger (3.11). They float freely over the cornea on the natural tears of the wearer (3.12).

Scleral-corneal contact lenses, which are easily visible on the eye, are not used routinely today. However, they may be prescribed for cosmetic purposes to cover an unsightly corneal scar. In this instance an artificial eye is painted on the

Fig 3.11

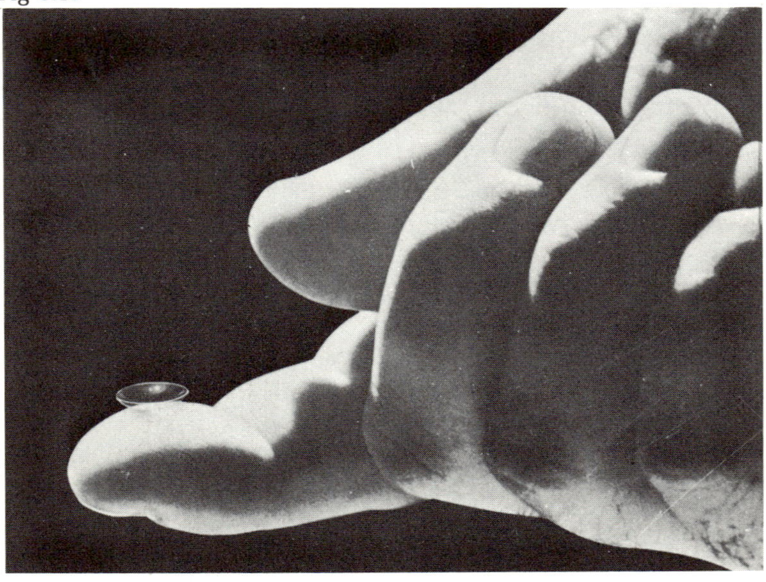

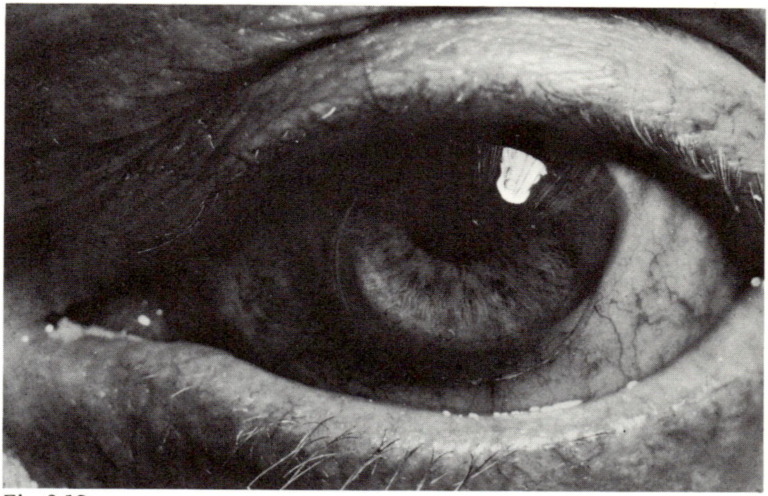

Fig 3.12

front surface of the contact lens. The scleral-corneal lens may also be used after heat or chemical burns or chronic infections to prevent scar formation between the lid and the cornea. They also may be prescribed to prevent exposure of the cornea when the lids have become paralyzed or when the corneal contact lens offers insufficient protection in vigorous contact sports. Neither comfort nor wearing time with the scleral-corneal lens is as great as that experienced with the smaller corneal contact lens.

Soft hydrophilic contact lenses, which are often difficult to detect in the eye, are more comfortable to wear than the small plastic lens. However, after six months to a year, visual acuity is reduced as the lens begins to deteriorate. At the present writing visual results with the new hydrophilic lens are not as good as those with the small plastic lens. Since the hydrophilic lens absorbs water, it may prove to be highly beneficial in patients with certain diseases of the cornea and eye.

Who benefits from contact lenses? If forward glasses cannot correct a visual defect, contact lenses may be prescribed. Persons who have had a cataract removed from one eye and have good vision remaining in the other eye often make excellent candidates for contact lenses. A person with imbalanced refractive errors that produce a marked difference in the image size of the two eyes (anisometropia) also may benefit remarkably. Wearing contact lenses equalizes the size of the images, enabling the two eyes to work together. Persons with a high degree of astigmatism, hyperopia, or myopia can avoid wearing thick glasses by choosing contact lenses. Persons with cone-shaped corneas (keratoconus), or those with scarred or irregular corneas benefit greatly because contact lenses form a smooth refractive surface. Persons with ear or nose defects or loss of ears also may be served well by contact lenses.

Contact lenses are worn for cosmetic purposes by actors, models, and some rather vain individuals. Since the lens may be colored, a blue-eyed person may appear brown eyed if he wears tan lenses, and vice versa. Tennis, basketball, or football players who fear that glasses will be broken or knocked off during the course of action use these lenses, as do swimmers or skin divers, to obtain clear vision.

Who does not benefit from contact lenses? Nervous, apprehensive, or sensitive persons may not tolerate the sensation of these "foreign bodies" in their eyes, while persons with tremor or severe arthritis may have difficulty inserting the lenses. Persons with a very small refractive error requiring a weak prescription for glasses are usually not motivated to wear contacts because of their high cost and the inconvenience of inserting them. Refractive errors not due to corneal curvature may not respond well to contact lenses. If a person has very poor sight in both eyes, he may not be able to see well enough to insert the lenses; with practice, however, he can overcome this obstacle. Some people do not see as well with contact lenses as they do with spectacles. Persons who are subject to recurrent infections in the anterior portion of their eyes and those with insufficient tears are discouraged from obtaining these lenses.

Misconceptions about contact lenses. *Wearing contact*

lenses will "cure" faulty vision so that a person eventually can discard his glasses entirely. This is incorrect. Contact lenses act like forward glasses. They correct, but do not cure, eye defects. Any improvement in vision is only temporary.

The need for eye examinations and changes of glasses is less frequent with contact lenses. This is not altogether true since the change of lenses is related to the degree of development of the eye or to crystalline lens changes. The contact lens will neither prevent a nearsighted eye from getting longer or more myopic, nor make a farsighted eye longer or less hyperopic. In some patients, however, nearsightedness does not progress after wearing contact lenses because of the effect of the contact lens on the cornea. Although it is true that the prescription for contact lenses does not have to be changed as often as that for spectacles and that contact lenses sometimes last for years, periodic eye examinations should not be abandoned.

Contact lenses are damaging to the eyes. Only if lenses are made poorly and are improperly fitted can they be injurious to the eye. Poor habits of the wearer may also cause untoward effects. If contact lenses are fitted and worn properly, they will *not* damage the eyesight or cause blindness.

Fitting contact lenses. The curvature of the corneal surface is measured accurately with a keratometer. Utilizing reflection patterns, this instrument reveals the power required to correct the refractive error by determining the proper corneal curvature. A photographic record of the corneal contact lens fitting may be made with a photoelectronic keratoscope.

Adjusting to contact lenses. In adjusting to the wearing of contact lenses the patient must follow rigidly the instructions of both his lens technician and his eye doctor. After the contact lenses have been fitted by the technician, all patients should return to their eye doctor within seven to ten days. At this contact lens examination, the eyes are refracted over the contact lenses to see if the lenses have been ground to the correct prescription, and the eyes are examined to see if the contact lenses have been fitted properly on the cornea. If the lenses are fitted improperly or if the prescription is incorrect, changes should be made immediately. During the adjustment period with contact lenses, the patient should plan on making repeated visits to his technician and eye doctor until an exact fitting is obtained. The patient's eyes may burn, tear, or itch, but these symptoms are usually minor irritations that may be relieved by eye medication.

Inserting and removing contact lenses. Before inserting the lenses, the patient always washes his hands with soap and water, rinsing thoroughly to be sure no soap residue is left on his hands and fingers. He then takes a contact lens between his thumb and index finger, and applies a few drops of some cleansing or antiseptic solution to the lens. The solution is thoroughly rinsed off with tap water, with the stopper down in the sink so that the lens will not accidentally go down the drain. With the contact lens on the tip of his most dexterous finger, he puts a drop of water or wetting solution inside the concave cupped surface of the lens. As he inserts the lens, the wearer separates his lids either by using the index finger and thumb of his free hand or by using both hands (3.13). He then places the lens on the cornea where it easily attaches itself because of surface tension between the fluid in the concave surface of the lens and the cornea. This makes insertion easy and discourages the wearer from jamming the lens into his eye. The lens should be inserted slowly either with or without a mirror.

Another way to insert the lens is to pull the lower lid down and then to insert the lens on the inside of the lower lid with the concave lens surface facing the cornea. As the wearer raises his lid, the contact lens falls into place on the cornea.

Plastic contact lenses are occasionally moistened by saliva or by applying them to the tongue. The only time this should be done is if fluid or water is not readily accessible

Fig 3.13

and the wearer has accidentally dropped the lens. At all other times, it is best to use the standard wetting solution. Hydrophilic lenses should never be moistened with saliva.

Men should insert their contact lenses before shaving. This will prevent beard particles—which may cling to their fingers—from adhering to the contact lens while they are cleaning them.

Lenses may be removed either manually or with a rubber suction cup. With the manual technique, the wearer can easily "pop" the lens out of his eye. Keeping his eye wide open, he places his fingertip at the outer corner of his eye and pulls the upper lid out and up toward his temple. His free hand is cupped awaiting the lens. This procedure may be carried out more safely over a half-full sink of water and more easily if the patient bends over. After the lenses have been removed, they should be stored promptly in their plastic case. Since the lenses themselves are plastic, some physicians recommend that they be stored in a water or antiseptic solution to keep them moist, as false teeth are soaked to keep them from drying out. If these precautions are not taken, the contact lens may warp and lose enough of its shape to interfere with vision.

Wearing time. If a patient has built up a tolerance of 4, 8, or 16 hours of wearing time with his contact lenses and then decides to go back to his glasses, he should not become disturbed if he cannot see well at first. Since contact lenses temporarily flatten the cornea to a small degree, the patient's particular refractive error may be changed (if he is myopic, he will be temporarily less myopic). After an hour or two of wearing glasses, his vision will return to normal. If the patient begins to wear his glasses after he has already established a tolerance of 14 or 16 hours with contact lenses, he should not go back to the 14 or 16 hour wearing time right away. The contact lenses should be worn for about six or eight hours until he rebuilds his level of tolerance. Contact lenses and glasses may be interchanged when the wearing time is up to ten hours. In fact contact lenses should be removed for an hour during the day, perhaps at suppertime, to allow the cornea to rest and to retain good vision with glasses. However, if the lenses are worn for too long a time after having used glasses, even though a tolerance has been established, the cornea may cloud or steam. Severe discomfort or a scratching sensation on blinking may indicate that the wearer has injured the cornea.

Danger signals. Wearers of contact lenses should be alert to these danger signals:

- Excess tearing.
- Cloudy, smoky, or foggy vision, which does not disappear upon blinking, while wearing the lenses.
- Excessive burning of the eyes.
- A scratchy sensation like that caused by a foreign body.
- Continuous redness of the eyes.
- Pain or discomfort while wearing or after removing the lenses.

If any of these signs or symptoms develops, the lens

should be removed and the closed eye patched with a sterile eye pad or handkerchief. If the symptoms disappear within 24 hours, the contact lenses may be reinserted. If the above symptoms persist or there is still a scratchy sensation in the eye, the wearer should consult his ophthalmologist immediately because he may have a corneal abrasion or infection. Any corneal lesion produced by a contact lens can be treated and cured if seen promptly by an ophthalmologist.

Complications from wearing contact lenses. Contact lens wearers may develop conjunctivitis, or they may sustain a corneal abrasion that could progress to a corneal ulcer. Abrasions may be caused by pressure of the contact lens against the cornea, by rubbing the eye, or by jamming the lens into the eye on insertion. Good hygienic measures will prevent the development of conjunctivitis. Inserting the lens slowly and remembering not to rub the eyes with or without the lenses in place will probably insure against abrasions. Contact lenses should be removed before going to sleep. The wearer may forget that the contact lenses are in place; upon awakening in the morning or in the middle of the night, he may try to rub the sleep from his eyes and produce a corneal abrasion. Sleeping with the lenses in place may also cause a corneal breakdown due to the lack of tear nutrition (overwearing syndrome).

Hints for contact lens wearers. *If you are going on a trip, always take your glasses.* You could develop conjunctivitis, or you could lose a contact lens.

If the wind is blowing strongly or dust is whipped up by a passing car, close your eyes. Turn your head or put your hands in front of your eyes to block out any foreign material. Remember that contact lenses do not provide the same protection as spectacles, which help shield the eyes from flying foreign objects. If you feel a scratchy sensation in your eye or if your eye is painful or irritated, a foreign body may have lodged between the cornea and the lens. Take the following steps: (1) *Do not rub your eye.* (2) Either gently pull the lower lid down, or the upper lid up. (3) Stick a finger into the inner corner of the eye on the lower lid and press the lid gently against the eyeball while looking up; this will tend to tilt the contact lens forward and allow tears to get between the contact lens and the cornea, helping to wash the foreign body away. (4) If the above procedures do not work, take the contact lens out and reinsert it. (5) If a scratchy sensation persists, see an ophthalmologist.

If you are playing football or soccer, squint your eyes slightly. Any pressure on the outer corners of the eye could dislodge the contact lenses, and squinting helps to discourage the lens from accidentally being jarred out of the eye. Do not swim under water or dive with your eyes open and your contacts in. The force of the water pressure might cause the lens to be dislodged and washed out. Instead dive and swim under water with your eyes closed. As you come to the surface, first shake your head to remove the excess water and then open your eyes.

Blinking more often, to help provide tear circulation when wearing your lenses, is not necessary.

Keep your head in a normal position when wearing your lenses. You do not have to look down or hold your chin up in the air to keep the lenses from falling out. The lenses float on the surface of the cornea on the surface tension of tears and eye secretions. If your lenses are fitted properly, they will not fall out.

Have your eyes examined one year after beginning to wear contact lenses. At this time the contact lenses should be checked for scratches which may easily and inexpensively be removed by polishing.

Bifocal contact lenses are frequently requested by patients, but at the present time they are not satisfactory. The person usually never sees as well with a bifocal contact lens as he does with his forward bifocal glasses. Although several types of these lenses are currently available, their proficiency has not been perfected. The patient may, however, wear contact lenses for distance vision and use a reading glass for close work.

SPECIAL LENSES

Various forms of special lenses are available to improve abnormal vision. For detailed close work, powerful **magnifying glasses** are available. These optical aids are similar to the old Sherlock Holmes magnifier or magnifying glasses set over the paper (3.14). Convex plus lenses produce magnification and a larger image on the retina. However, the stronger the lens, the closer it must be held to the object for clear focus. The statement that optical aids will hurt the eye is absolutely false; their only purpose is to magnify objects for better sight.

Fig 3.14

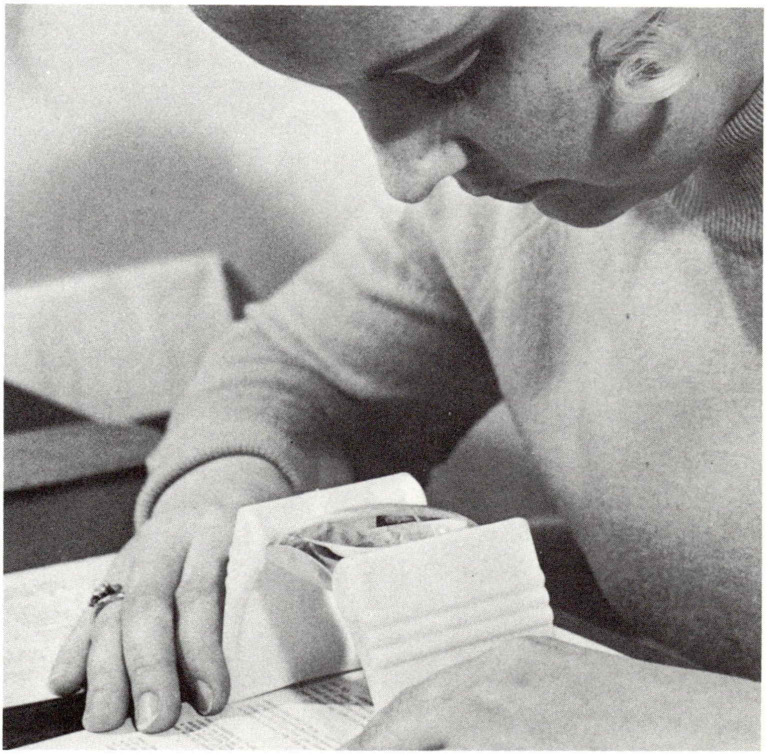

For people with marked visual impairment caused by disease within the eye, **telescopic** or **spherical conoid lenses** are helpful for handwork, sewing, or reading (3.15). Although telescopic lenses, like binoculars, are used to help bring distant objects close, only one eye can be used at a time for close work. Telescopic contact lenses are also available.

Cataract lenses no longer have to be heavy, as they can now be made of plastic instead of glass. Special conoid or aspherical cataract lenses now permit vision through the entire lens, instead of just through the center.

Prismatic spectacles permit a person lying flat on his back to read comfortably. They also enable a person giving a speech to read his notes while he holds his head erect. Although he appears to be looking intently at his audience, he is gazing at his notes.

EXAMINATION OF VISUAL FIELDS

The field of vision embraces the eye's total area of sight. Normally objects above and below as well as to the left and right of the central line of vision can be seen without turning the head. Tests to determine the extent and to localize the cause of visual field loss are called visual fields or perimetric studies. Two types of tests are used.

Testing the Central Visual Field

The function of the central seeing portion of the eye is tested by using a tangent screen or some similar instrument to measure the sight from 0 to 30° from the central point of fixation (3.16). The patient sits with his chin in a chin rest, facing the screen or instrument, which may be placed from 0.33 to 2 meters away from the eye depending upon the instrument used. Each eye is tested separately with the other eye covered. While the patient looks at a white central fixating target, the examiner brings in white or colored test ob-

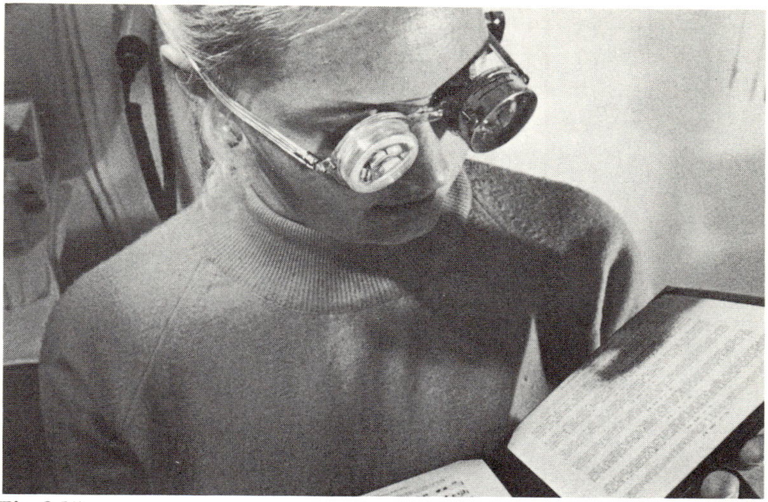

Fig 3.15

jects of various standard sizes from the outside of the screen. The patient watches the central point of fixation and, out of the corner of his eye, notes the first appearance of the test object. This result is then recorded for each meridian, direction or angle, of the test field of vision.

Fig 3.16

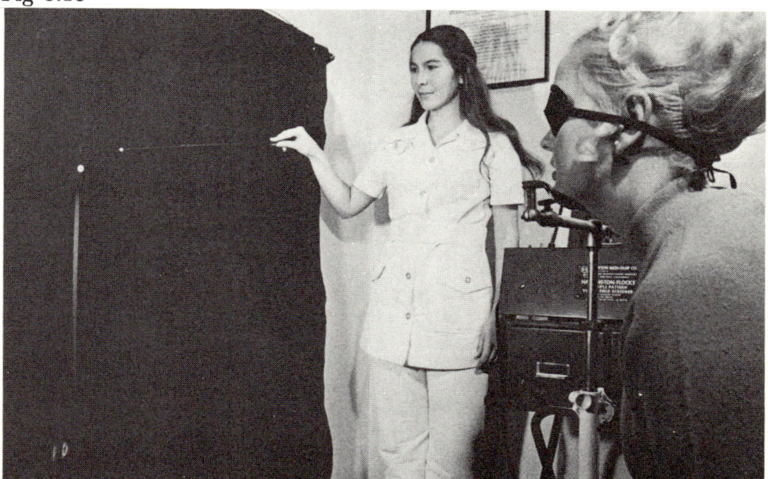

When vision is tested at the point where the optic nerve leaves the eye, light or the image of the stimulating object is not seen. This blind spot (scotoma), which is due to the absence of overlying retina at the optic nerve's point of exit from the eye, is a normal finding in all persons. No one is aware of its presence. The spot is tiny and sight in that area is covered by the overlapping visual field of the other eye.

The blind spot can be detected by marking a dot on a piece of paper and then putting an "X" one to two inches away from the dot. Now close one eye and look at the dot while moving the paper back and forth slowly. At some point the "X" will disappear. The point of disappearance is the blind spot. Abnormal areas of blindness detected during the central visual field studies are called scotomata.

Testing the Peripheral Visual Field

The side vision of the eye is tested with a perimeter or arc-shaped instrument (3.17). Although this test is conducted

Fig 3.17

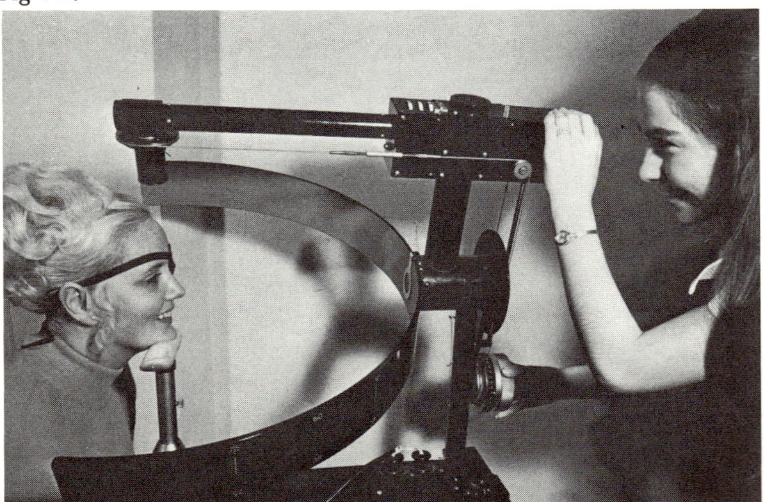

in a manner similar to the test for central vision, the field of sight is measured up to 90° from the point of fixation. The appearance of the test object is gauged for the various meridians, and the final field of sight is recorded. Normally a person should see about 60° toward the nose, 50° upward, 90° temporally, and 70° downward from the central point of fixation.

Confrontation Field Studies

When central and peripheral visual field testing equipment is not available, a simple but much less accurate test of the visual fields can be made by confrontation field studies. When the right eye is tested, the patient covers his left eye with his hand and the examiner closes his own right eye. With the patient's right eye looking directly into the examiner's left eye, the examiner brings a pencil or bead or even his finger from the various meridional fields of vision —up, down, left, and right—toward the central fixation point. He then determines if the patient observes the incoming object at the same time that he does. If this occurs, and assuming that the examiner has a normal field of vision, the patient's visual field is also normal. The same procedure is then performed on the left eye.

FUNDUS EXAMINATION

During a complete eye examination, the entire fundus is studied through a dilated pupil with an ophthalmoscope (1.4 and 1.5). This instrument directs a beam of light into the patient's eye. After the light passes through the pupil, lens, vitreous, and on to the retina, it is then reflected back through a hole in the ophthalmoscope head out to the doctor's retina. Here it is brought into a sharp focus by a series of lenses on a circular rotating disk in the ophthalmoscope head. Since this adjusts for any refractive error in the patient's or doctor's eye, the fundus can be viewed more accurately. This particular instrument is used for direct ophthalmoscopy. The details of the fundus are seen uniocularly (with one eye) by the doctor (1.5). With a binocular ophthalmoscope, he can use both eyes at the same time, bringing depth perception to the direct ophthalmoscopic examination.

Indirect ophthalmoscopy, which is also binocular, is performed by directing a light into the patient's eye through a strong spherical lens held a short distance away (3.18). This permits the doctor to visualize the back of the eye. Recently a new uniocular indirect ophthalmoscope, which looks much like an ordinary head ophthalmoscope, has been made available.

Fig 3.18

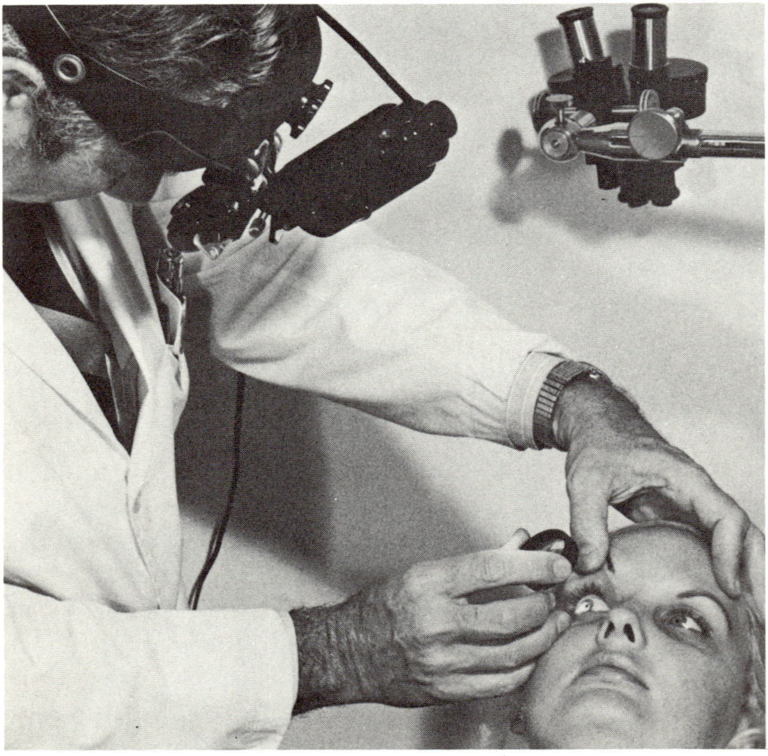

Regardless of the type of ophthalmoscopy used, the signs of many systemic diseases are often first discovered by this study of the fundus. It is the only place in the body where blood vessels first can be studied by the naked eye and then photographed (3.19). Here hemorrhages, tumors, and other tissue changes may become apparent. Arteriosclerosis, hypertension, toxemia of pregnancy, nephritis (kidney disease), heart disease, diabetes, tuberculosis, syphilis, anemia, leukemia, cancer (metastatic tumors from elsewhere in the body), brain tumors, or abscesses may be discovered.

Eye Signs and General Disease

The first signs of many other general conditions may be exhibited by early eye changes. Multiple sclerosis, brain tumors, or hemorrhages (3.20) may cause double vision. A muscle imbalance caused by paralysis of one or more eye muscles produces internal or external strabismus. Thyroid disease or lesions of the brain may cause protrusion of the eyes. Myasthenia gravis may be detected by drooping of the eye-

Fig 3.19

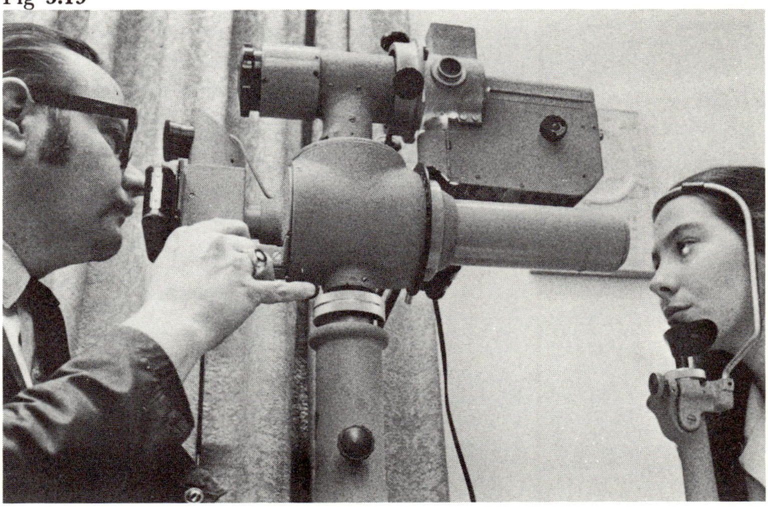

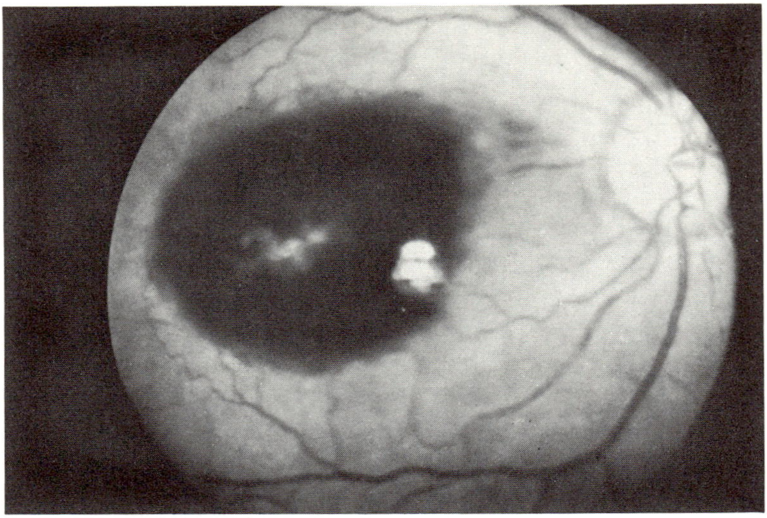

Fig 3.20

lids. Vitamin deficiencies can instigate disease in the front of the eye or night blindness. Liver disease may be indicated by a jaundiced or yellowish color of the eyes or skin called icterus. Infected sinuses or teeth may produce referred inflammation or pain in the eyes. Iritis and chorioretinitis, which are inflammations in the eye, may be secondary to syphilis, sarcoidosis, or tuberculosis. Kidney disease, thyroid disease, or allergies may cause the lids to swell.

Periodic eye examinations by an eye specialist may lead to the detection and diagnosis of these systemic diseases. In some instances only through finding and treating these diseases can the eye condition be cured.

A PRIVATE CONSULTATION ON YOUR EYES

Author Abrahamson offers some practical advice on eye care . . . not only for you but for your youngsters.

MEDCOM: Are Americans taking care of their eyes?

ABRAHAMSON: I think Americans are taking better care of their eyes than are people in other countries. We have good eye physicians—ophthalmologists giving them excellent care. The main thing is to get the patient to the doctor.

Why is that a problem?

Some people take for granted that if they see well they don't have to do anything about their eyes. But I feel an ounce of prevention is worth a pound of cure.

What kind of prevention do you recommend?

If there's history of crossed eyes in the family, youngsters should be examined sooner than usual, even at the age of 2, so that if they have a defect it can be picked up and corrected before it's too late.

What's the first indication that the child has a problem?

The child may squint. He may have a deviating eye, a crossed

I recommend that all adults have a routine eye examination every one to three years. Children should be examined before the age of 6.

eye, or a walleye. He may have to get too close to the television to see it. He may rub his eyes.

Do most of these things appear in the home first? Or do they appear at school?
They can appear both at home and at school. But the main thing is to make sure that every child has a routine eye examination, where the pupils are dilated with drops or ointments called cycloplegics. This helps rule out not only the need for glasses, but disease in back of the eye.

What are the most frequent problems kids have with their eyes?
The most common problems are refractive errors, like near- and farsightedness, and strabismus or squint. Many of them have infections in their eyes, such as a conjunctivitis or blepharitis, which is inflammation of the lining of the eye or eyelids.

At what age are these problems likely to . . .
. . . any time from infancy to 15 years, depending on the child.

Why do you recommend annual physicals for children . . . in terms of their eyes?
That's a must, because if a child is seen before he's 6 years old, a defect can be diagnosed and discovered and can be corrected—most can be corrected. The main thing is making the diagnosis, just the same as in an adult.

Is there any kind of do-it-yourself eye test that parents could conduct on their kids, or do you recommend that they go right to an ophthalmologist?
There are eye tests that parents can do on their children. They could test them to see what objects they can see across the room, comparing one eye to the other. The school test

program that they have is like this, where the child looks at a chart. They cover one eye and then the other just to see what the child sees. The only trouble is if they don't completely cover up a normal eye, the child may be seeing 20/20 with his good eye as well as appearing to have 20/20 vision in his blind or partially seeing eye. So the technique for taking the vision is very important. Even by taking the vision a child could have 20/20 vision in each eye and still have a defect that could go undiscovered unless he was seen by a specialist, such as an ophthalmologist.

Should children wear contacts?

No one should be forced to wear contact lenses. They should want to wear them. Contact lenses are similar to glasses except they're worn on the eyeball, and a person should have the innate desire to wear the contact lens. If they don't want to wear them, you can't insist. Contact lenses replace glasses, but they don't cure eye disease.

Contacts should not be worn unless a person has an innate desire to wear them.

Is it true that in some so-called high pollution areas you have trouble wearing contact lenses?
Yes. Contact lenses are a foreign body in the eye and may irritate it, but where pollution is in the air it's even worse. It's exaggerated.

Can anybody wear contact lenses, or do some people perhaps congenitally have a problem wearing them?
Some people may have a problem wearing them, especially if they have dry eyes. They may not have enough tear secretion. Different refractive errors also prohibit some from wearing contact lenses . . . for instance, if they have a huge keratoconus, or cone-shaped cornea. Also, if a person has recurrent infections in the eye he should not wear contact lenses.

Now, new subject—sunglasses. Do you recommend them?
If someone is sensitive to bright lights, yes, sunglasses are excellent.

How do you know if you're sensitive to bright light?
You can tell if your eyes are uncomfortable when you're out in bright light . . . if too much glare bothers you. Some people have light sensitivity. This usually occurs in people who are light complected and have blue irides.

What about these new, so-called self-adjusting lenses that seem to change shades from indoors to outdoors?
The glass you're referring to is a photo-gray lens—a wonderful invention. It's due to the chemical materials in the lens itself, and they are excellent because in bright lights the pigment absorbs the light and turns gray and in a darker atmosphere the pigment disappears and the lens becomes clearer.

Are you wearing that kind right now?
Yes, because I like them for golf and for indoors—you only need one pair of glasses for both. However, in the bright sun-

light the lenses may not be dark enough for some people. Their only drawback is that it takes a little while for the lens to go from dark to light.

How long?
It could be up to 20 or 25 minutes. And so people must be very careful when they're, say, driving through a tunnel . . . especially if they've been out in bright sunlight and the lens is really dark.

Everybody talks about 20/20 vision, but I'll bet if you went up to 30 people on the street outside your office they couldn't tell you what it means.
Well, I think everybody knows that 20/20 vision is normal sight. It means that when a person is standing 20 feet away from a chart he will be able to see the 20/20 line on the chart. If a person can see *only* the big E at 20 feet and cannot read any of the other lines, he has 20/200 vision. On the other hand, a person with normal vision (20/20) could read the big E at 200 feet.

Okay. Other often misunderstood terms are nearsighted and farsighted, mainly because they seem ambiguous. If you're nearsighted you can't see things far away. But if you can see things far away then you're farsighted.
Well, nearsighted means just what it says. You are near-sighted. You can see things up close. And therefore you cannot see things far away. Farsighted means what it says. You can see things far away but you may have difficulty seeing close-up. It's just terminology. Nearsightedness is corrected with a myopic or concave lens, and farsightedness is corrected with a convex lens, which focuses the ray of light forward from behind the retina to the retina itself. And in myopia, which is characterized by an elongated eyeball, the concave lens focuses the rays that are normally in front of the retina back to the retina.

While you're talking about retinas, I hear a lot about detached retinas. How do they happen and how serious are they?

Detached retinas can be a very serious problem. The retina is the seeing part of the eye. It's like the film of a camera. And if the retina becomes separated from the back of the eye, from the choroid, it can lose its function. If it loses its function, the patient doesn't see. Many cases result from an adhesion between the vitreous and the retina. The vitreous of the eye will pull on this adhesion, causing the formation of a hole. The vitreous fluid, which is like water, gets through this hole and pushes the retina away from the choroid . . . as if you were to pour water through a hole in wallpaper causing the water to precipitate downward and push the wallpaper away from the wall.

Any other causes?

Another common cause is trauma. People, especially athletes, can receive blows to the eyes. That can produce a detached retina. People who are very nearsighted have an elongated eyeball and they're subject to the possibility of a detached retina. Shrinkage of the vitreous in the eyeball can produce a detached retina. Many patients inherit congenital weak spots in the eye.

Somebody once told me that even a sneeze could cause a detached retina. Is that right?

It's possible but unlikely that a sneeze could be so severe that the pressure in the eye is built up enough to force a detached retina. There are many causes. But today we have good techniques for repairing the detached retina—fitting it back in position. Many times vision can be restored to normal.

Would you say the rule, if you suspect a detached retina, is to see . . .

. . . an ophthalmologist immediately. Now ophthalmology is

so specialized that we have colleagues who do nothing but retinal detachment surgery.

We also often hear that cataracts are a problem. In fact, a fellow in our office just had a cataract operation the other day. Can you briefly describe what happens and what the problem is with them?

Well, cataracts are a clouding or opacity of the eye lens which is located in the black pupillary space. I always compare it to a window. If the window is clean, everything on the other side is clear. If you put soap on the window and you look out, everything's smoky, or cloudy or fuzzy, the way it looks to a person who has a cataract. Normally you can take a rag and wipe the soap off the window, but if you put paint on the window you can't wipe it off. No matter how many clear glasses you put in front of your eyes, you're not going to see out any better as long as that painted glass is there. So if you want to see better, you remove the painted glass window and replace it with a clear, clean glass. A cataract is like paint be-

No one should ever fear having a cataract operation.

ing smeared on the lens of the eye. If it's there, it can be operated upon and removed surgically, painlessly. Sight can be restored if the back of the eye is normal.

Are most cataract operations successful in that a person regains vision?
We have newer, modern surgical instruments, microscopes to see better, newer suture material . . . I would say over 98% of cataract surgery is successful today. If you have to call any operation pleasant, this is probably the most pleasant one you could have and it's the most rewarding because you make the blind see. It's really very dramatic, and the patients that walk in with a white cane can throw their canes away when the cataract is removed and they have the cataract glasses. In fact, I think it's the most exciting operation in all of medicine.

You've done a lot of these, haven't you?
Well, it's one of my specialties, and yes, I've done several thousand . . . very gratifying work.

Earlier you talked about annual eye examinations for children. Do you recommend the same for adults?
I recommend that all adults have a complete eye examination every one to three years; and if they have eye problems they should be seen even more often. The only way we're going to be able to diagnose disease is by seeing the patient and actually putting him through a thorough eye examination where the pupil is dilated, the back of the eye studied thoroughly, and the intraocular pressures for glaucoma taken routinely. Not only is it an examination for glasses and muscle balance, but also a complete examination.

Do you honestly think most people realize the importance of a thorough physical as you just described?
Many people take their eyes for granted, and if they can see well enough to get around they're happy. They don't want to

go to the expense of getting an eye examination, which actually is not expensive. It's probably one of the cheapest investments a person can make. People will have their heart listened to, they'll have their lungs listened to, they get an ECG, but their eyes they take for granted. And if you ever try tying a handkerchief around your eyes and walking around just one day without seeing, you'd appreciate your eyes.

Please be specific. What should your annual eye physical include?

It should include an examination of the external portion of the eye . . . the lids, the conjunctiva, the sclera, the cornea. the iris, the pupil. You should check the intraocular pressure in the eyeball for glaucoma and the ocular movements for motility. The doctor should do a funduscopic examination, looking at the back of the eye . . . the retina . . . to see if there's any disease there. Remember, the eye is the window of the brain and body. Many times kidney disease, diabetes,

At eye examination—which actually is not expensive—is probably one of the cheapest investments a person can make.

high blood pressure, arteriosclerosis, and brain tumors are diagnosed just through an eye examination.

Anything else?

If necessary, specialized tests—called provocative tests—are performed if a patient is suspected of having glaucoma, for instance. There are other tests, such as visual field studies, to rule out any defect in the pathway of vision going from the eye to the brain. You know, you don't see with the eye. The eye is just the receptor. You actually do your seeing with the brain. So the main thing is to have the patient seen, do a complete eye examination where we look for minute details with technical instruments such as a slip lamp, which is a high-powered light with a microscope. Through this type of examination disease can be discovered and corrected and the patient's eyes can be preserved.

4

PROBLEMS IN DIFFERENT PARTS OF THE EYE

MANY DISEASES AFFECT THE EYE. Here we shall discuss only the most commonly encountered and the most often misunderstood eye disorders. Photographs provide an intimate glimpse of disorders of each part of the eye and its surrounding structures (4.1).

LIDS

Ptosis

Ptosis, or drooping of the upper lids (4.2), may involve one or both eyes in children or adults. The upper lid hangs down over the cornea and sometimes interferes with vision. Although a crutch glass may be used to elevate the lid, surgery is often required to correct this defect. Ptosis present since birth may be accompanied by a paralysis of the superior rectus, the elevating eye muscle. When ptosis occurs in adults, it may be the result of a systemic disease, such as myasthenia gravis, which can be treated medically or surgically. Ptosis can also follow muscle or nerve damage in other parts of the body, or tumors of the lid. When ptosis occurs suddenly in one eye, disease of the brain itself must be considered, and the patient should be seen at once by a physician.

Fig 4.1

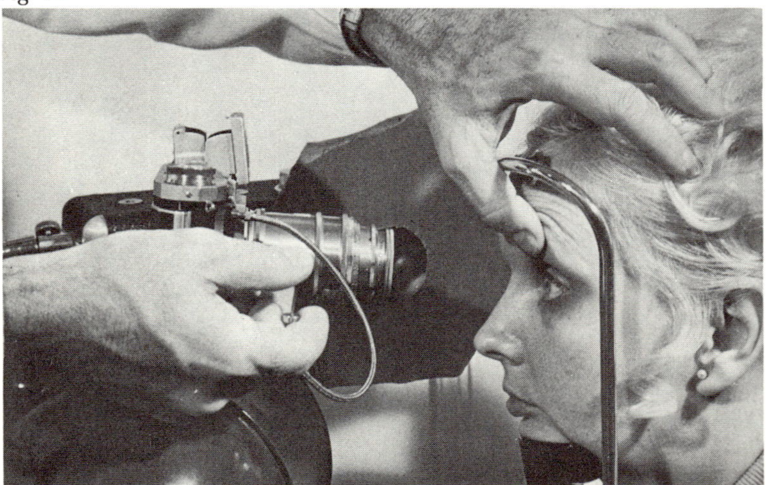

Blepharitis

This common infection of the lids of both children and adults is characterized by itching and redness accompanied by scales or crusts on the lashes, which may cause the lids to stick together in the morning (4.3). If the patient rubs his eyes for relief, as he usually does, he initiates a vicious cycle of itching, and secondary infection which follows. Blepharitis often is associated with dandruff or with scaling of the scalp, called seborrheic dermatitis; both will improve with medical treatment. Blepharoconjunctivitis is an inflammation of the conjunctiva and lid. Blepharodermatitis is an inflammation of the lid margin and overlying skin characterized by swelling, itching, and redness of the lids. This disorder can be an allergic reaction to drugs, poisoning, or infection. It responds to medical therapy. Contact dermatitis, caused by soaps or cosmetics, also results in inflammation of the lids. This irri-

Fig 4.2

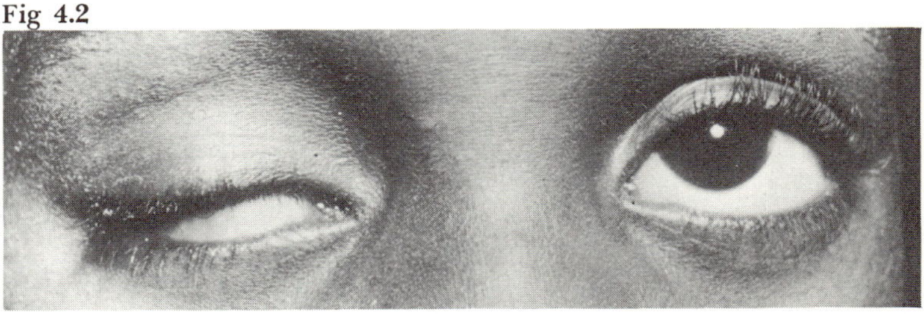

Fig 4.3

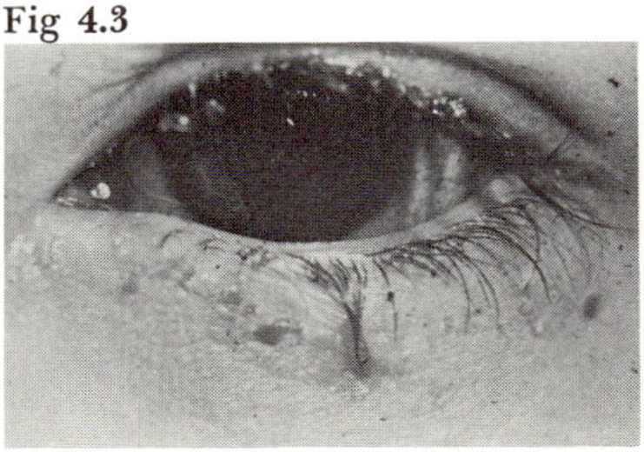

Fig 4.4

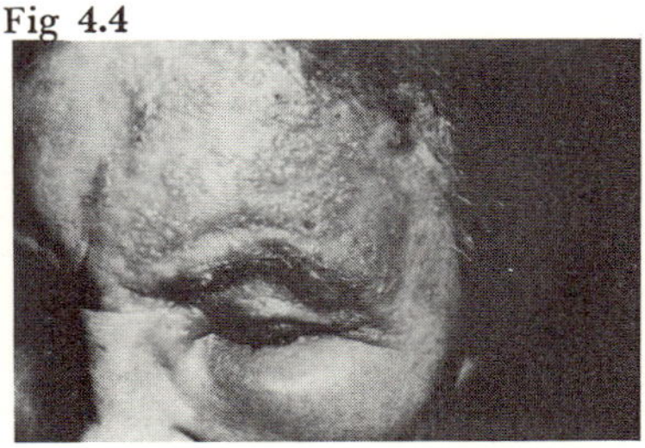

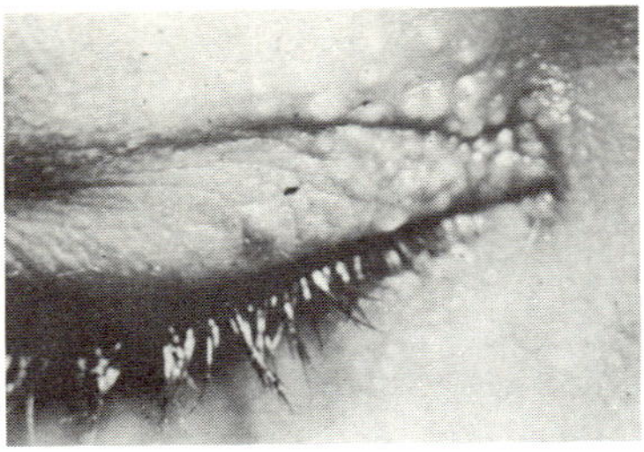

Fig 4.5

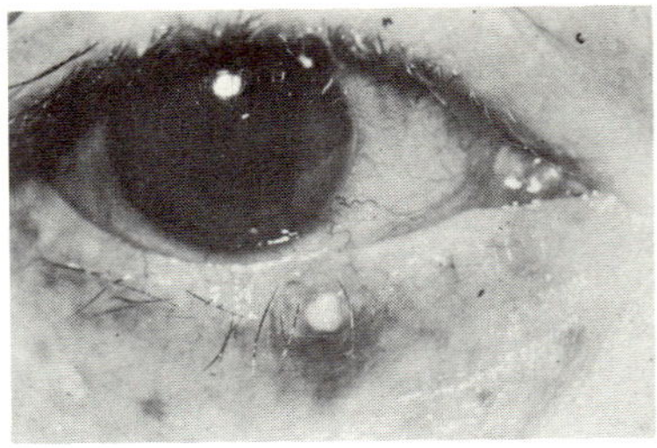

Fig 4.6

tation can be controlled by avoiding the sensitizing agents. Both herpes zoster (4.4) and herpes simplex (4.5) are infections caused by a virus. They can involve the lids and require medical therapy.

Sty and Chalazion

A sty, or hordeolum (4.6), is an infection involving one or many of the roots of hair follicles of the upper and lower eyelashes. If a refractive error or eyestrain leads a person to rub his eyes with contaminated fingers, he may embed an infectious organism in the roots of his lashes. The red, painful, elevated lesion is sometimes accompanied by swelling and inflammation of the lid. The sty also may develop a white center. If this is opened and drained, spontaneously or by surgery, the inflammation subsides and eventually clears completely.

A chalazion (4.7) is an accumulation of material in the meibomian gland. About 25 to 30 of these glands are located along the border of the upper and lower lids. Normally they secrete a material that helps to lubricate the eye so that the lid can move up and down smoothly. If an infection or allergic reaction blocks the duct of this gland, a chalazion is produced. It may be single or multiple and acute or chronic. An acute infected chalazion is often confused with a sty since it is red, elevated, and painful and may cause the entire lid

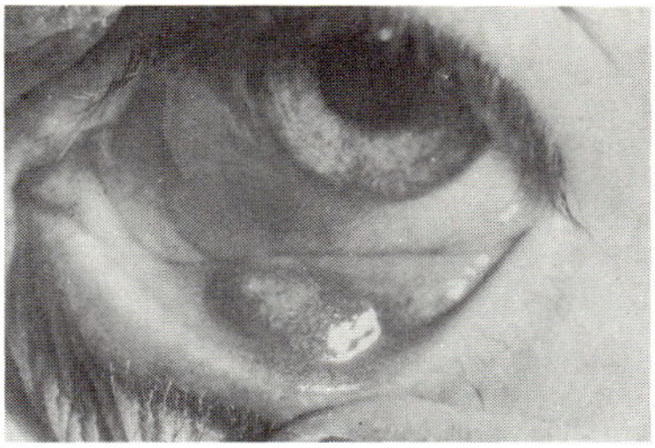
Fig 4.7

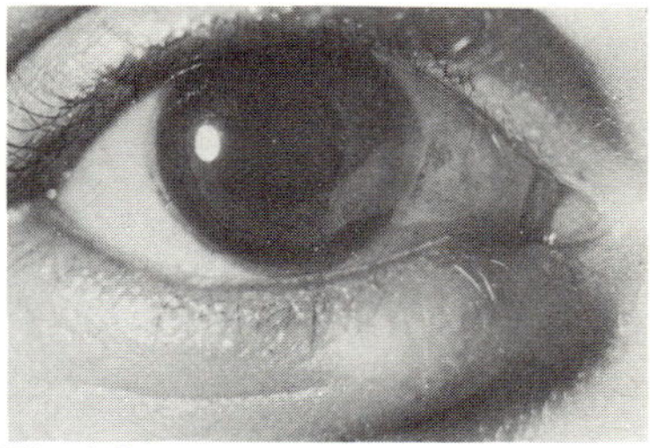
Fig 4.8

to swell. A chronic chalazion may appear as a painless lump on the lid and it may distort vision (4.8). Although sties may clear by themselves, chalazions usually require surgical opening to bring about a complete cure. Until a physician can be consulted, warm water compresses applied to the eye may provide much desired relief in acute cases.

Trichiasis

If eyelashes turn in toward the eyeball (4.9) and scratch the cornea, they produce a sensation like a foreign body. Trichiasis may result from trachoma, burns, or injuries to the lids. Removal of the offending lashes or plastic surgery performed on the lid relieves the symptoms.

Twitching of Lids

This annoying occurrence is often due to fatigue, emotional disturbance, or eyestrain from an uncorrected refractive error. Blepharospasm is a forceful spasmodic squeezing of the lid due to irritation in the eye or in the seventh cranial nerve, toxic drugs or "pep pills," psychologic reaction, or the same causes as for twitching. Rest, internal medication, or removal of the causative agent often will alleviate the problem.

Bags or Shadows Under Lower Lids

Although these shadows may be cosmetically unpleasant, they are not actually harmful. Blepharochalasis (4.10) is an abnormal relaxation of the excess skin of the upper or lower lid. With upper lid involvement skin may hang down and interfere with vision or be cosmetically unbecoming. This condition usually occurs in elderly people as the elastic tissue in the structures of the lid relaxes. Another cause may be the outward bulging of fat from the orbit through the protective membrane that normally separates the outside of the eye structures from the underlying orbital contents. In young adults fatigue, lack of rest, or congestion of the tissues might be considered likely causes. Plastic surgery is often helpful (4.11, 4.12).

Xanthelasma

This fatty yellow or orange tumor of the skin, sometimes appearing in a butterfly distribution, is located in the

Fig **4.9**

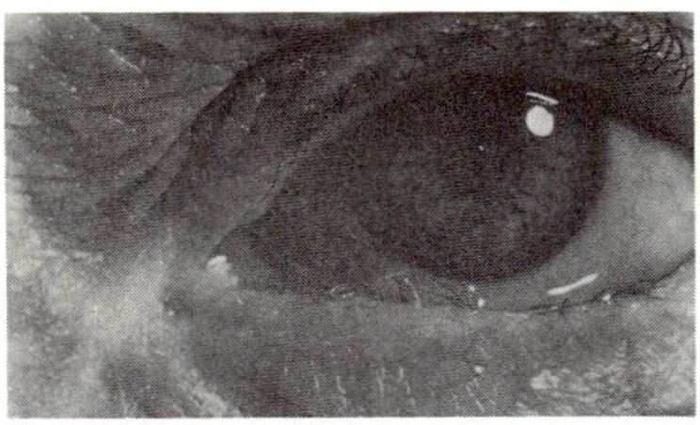

Fig **4.10**

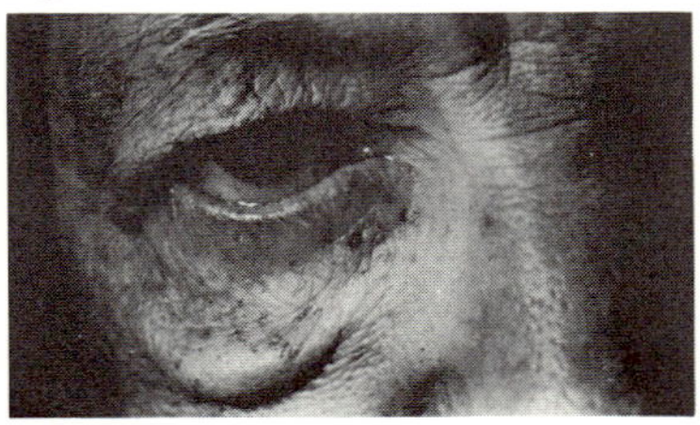

Fig **4.11**

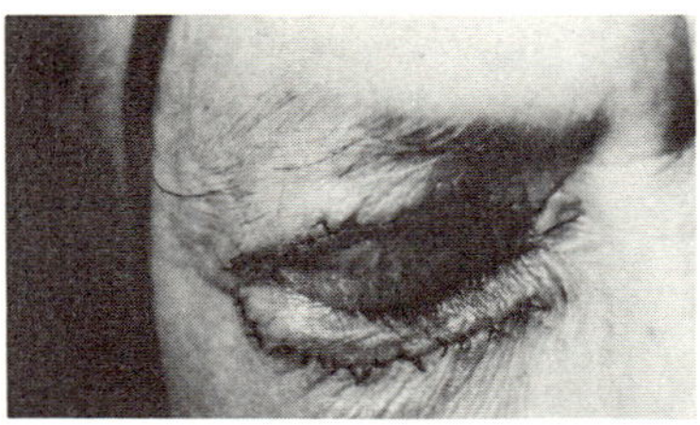

Fig **4.12**

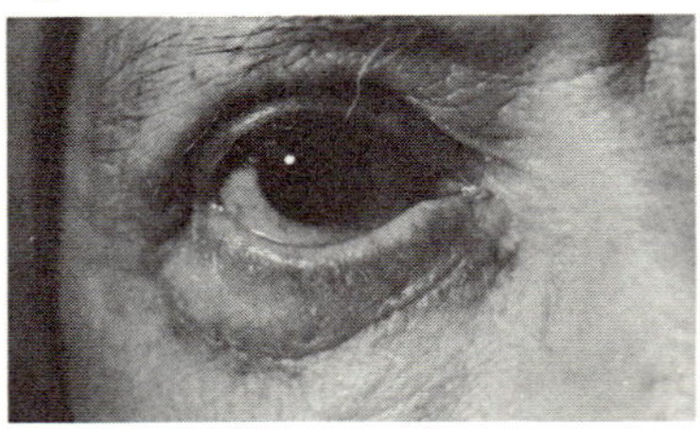

area near the nose on the upper and lower lids (4.13). It is not cancerous and will do no harm, but it does produce a blemish. In this instance it is removed painlessly by plastic surgery. Since it may be indicative of a high blood cholesterol or diabetes, persons with xanthelasma are advised to consult their general physician.

Tumors

Tumors on the eyelids or sores with scales that often crust and bleed may be serious and usually do not disappear by themselves. The early diagnosis of a cancerous lesion (4.14) should be entertained, but this can only be verified by biopsy or surgical removal of the lesion. Immediate treatment should not be delayed.

There are other skin tumors of a less severe nature. **Papillomas** (4.15) are elevated skin tumors. **Verrucas** are like warts. **Nevi** are pigmented tumors. Any of these tumors may involve the lids and may be removed for cosmetic, functional, or diagnostic purposes.

Fig **4.13**

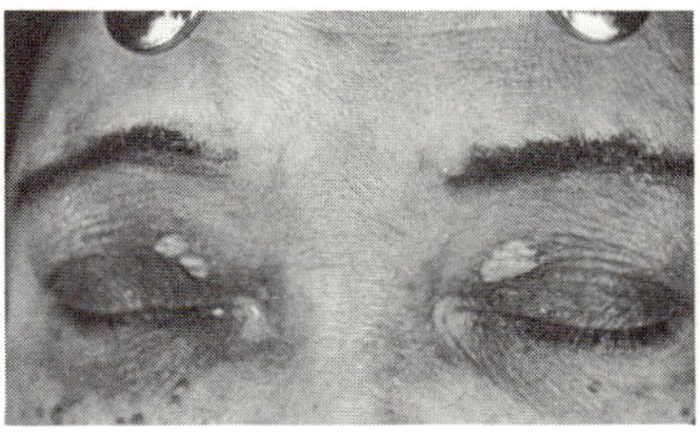

Fig **4.14**

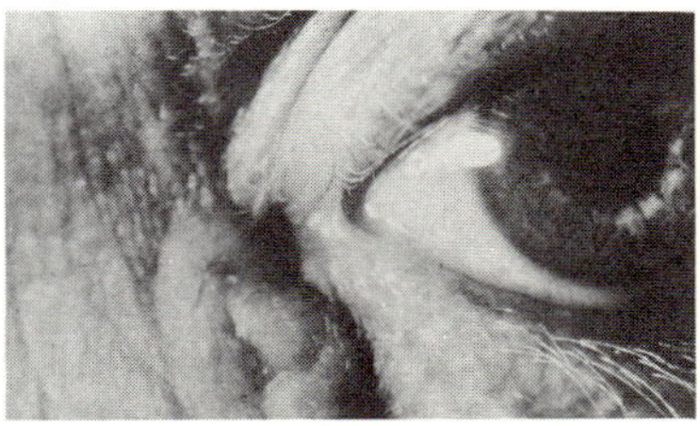

Fig **4.15**

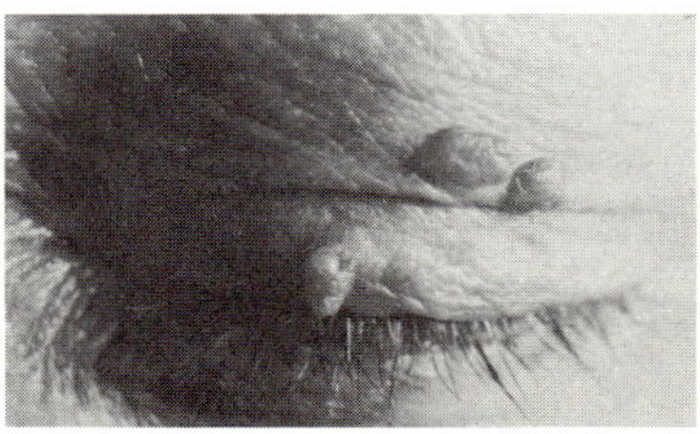

Fig **4.16**

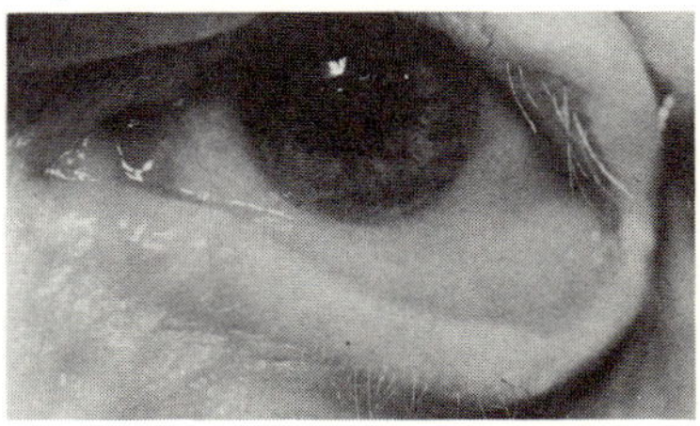

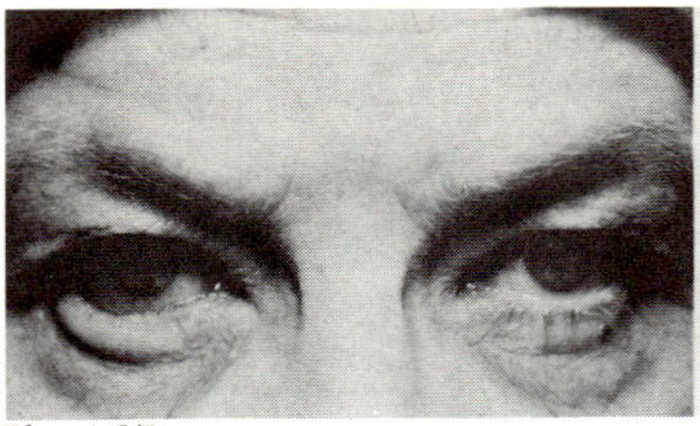
Fig 4.17

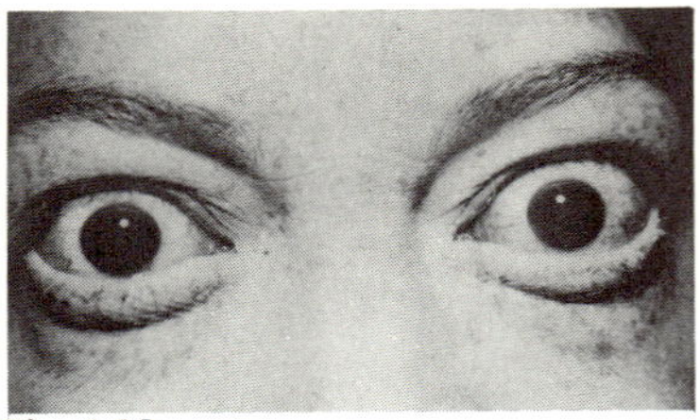
Fig 4.18

Ectropion

In this condition the lower lid usually turns away or everts from the eyeball (4.16). Ectropion may be due to laxity of the tissue in elderly people or to Bell's palsy (paralysis of the seventh cranial nerve), which causes weakness of the muscles of the lid. It may also follow cuts, infections, or burns of the lids and face that heal poorly; the resultant scar tissue forms adhesions that cause the lids to turn out. Besides being cosmetically unpleasant ectropion is accompanied by troublesome tearing and infection. Scotch® or adhesive tape may be used temporarily to hold the lid in place until corrective plastic surgery can be performed.

Entropion

Entropion is the opposite of ectropion. The lid is inverted or turned into the eye (4.17), causing the lashes to scratch the cornea and produce irritation. Tearing and secondary infection as well as an unpleasant looking eye cause the patient to seek medical care. Entropion may be the result of spasm or secondary contracture or strictures from burns, injury, or an infection such as trachoma. It may involve the upper or lower lids. Scotch® or adhesive tape applied to the skin of the lid temporarily may evert the lid and relieve the annoying symptoms. Corrective plastic surgery is usually required.

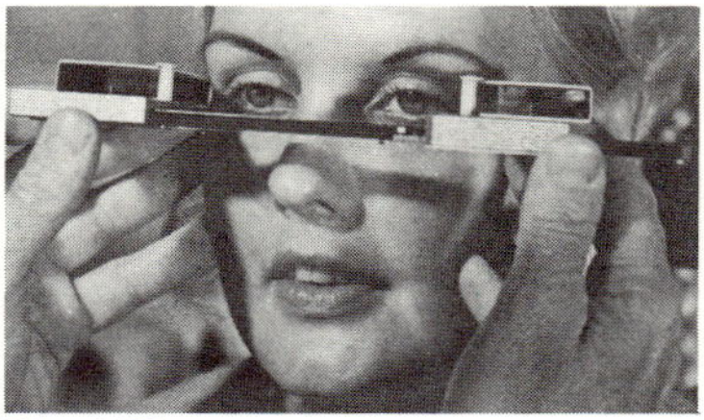

Fig 4.19

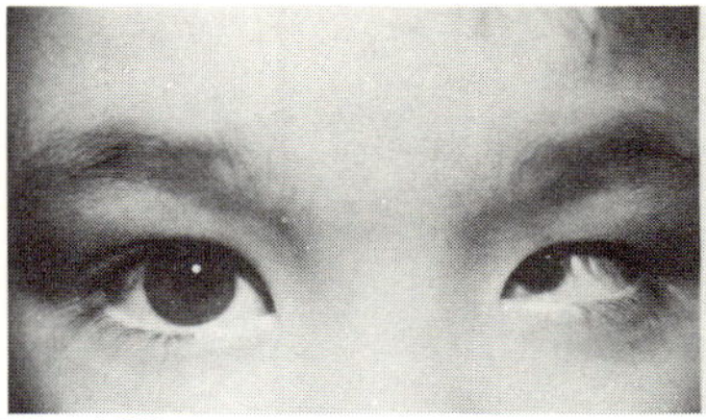

Fig 4.20

Lagophthalmos

Incomplete closure of the lid may be caused by scarring or weakness of the muscles of the lid. If the cornea is exposed, drying and inflammation may accompany this condition. Treatment is directed toward protecting the cornea from exposure by topical medication or plastic surgery.

Exophthalmos

The eyeball proptoses or protrudes from its socket, producing a widening of the palpebral fissure (space between the upper and lower lid). Since the patient does not blink frequently, he develops a staring gaze (4.18). The degree of protrusion can be measured with an exophthalmometer (4.19). Exophthalmos may be the result of: an endocrine disorder (thyroid disease); a tumor or such inflammations as cellulitis or periostitis behind the eye or in the nasal sinuses; or an aneurysm (a bulge in a weakened blood vessel wall) interfering with the blood flow from the eye to the brain. Medical and surgical care may be necessary to correct this condition.

Enophthalmos

In enophthalmos, which is the opposite of exophthalmos, the eye is sunken into the orbit (4.20). This abnormality may be present from birth, or it may be due to contraction of the eye muscles pulling the eye into the socket. Other causes may be absorption of orbital fat or an injury to the eye with resulting fracture of the orbit.

Symblepharon

With this condition a band of scar tissue extends from the lid to the eyeball (4.21) as a result of injury, chemical burns, or infection of the conjunctiva.

CONJUNCTIVA

Conjunctivitis

Inflammation of the conjunctiva is called conjunctivitis. It is accompanied by redness, itching, burning, tearing, or a discharge (4.22). The character of the discharge may vary according to the type of infection present. With diphtheria infection the discharge forms a membrane. Gonorrheal or

Fig **4.21**

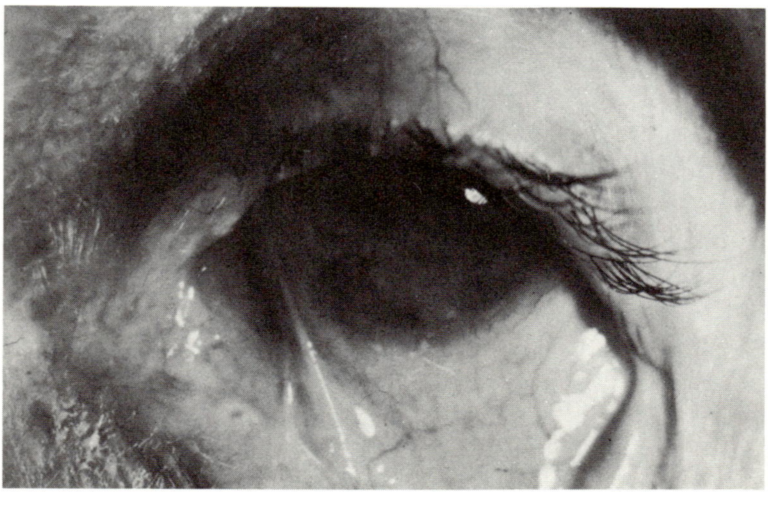

other types of bacterial conjunctivitis are accompanied by a pus-like discharge. Viral conjunctivitis has a watery discharge. When the conjunctiva becomes inflamed from such irritating agents as soap, the condition is called chemical conjunctivitis. Conjunctivitis may also be caused by wind, dust, or smoke. Pink eye is a form of conjunctivitis. Even though some forms of conjunctivitis are not contagious, patients should use their own towels and avoid rubbing the eyes to prevent possible contamination.

Among the several contagious forms of conjunctivitis are: infectious conjunctivitis, trachoma, and epidemic keratoconjunctivitis. Trachoma once was a major cause of blindness, but today it is rare in this country thanks to modern antibiotics and the sulfa drugs. It still flourishes, however, in the Middle and Far East. The word trachoma is often confused with the terms glaucoma (increased pressure within the eye) or glioma (tumor in the eye).

With catarrhal, allergic, purulent, traumatic, viral, follicular, or vernal conjunctivitis there may be a foreign body

Fig 4.22

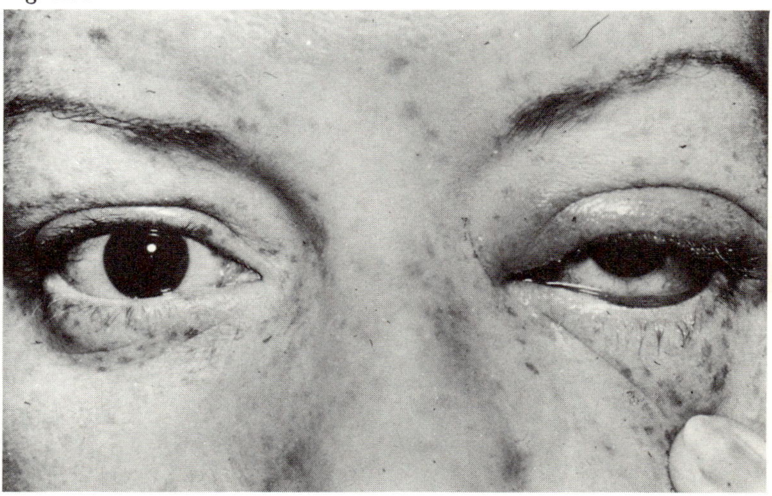

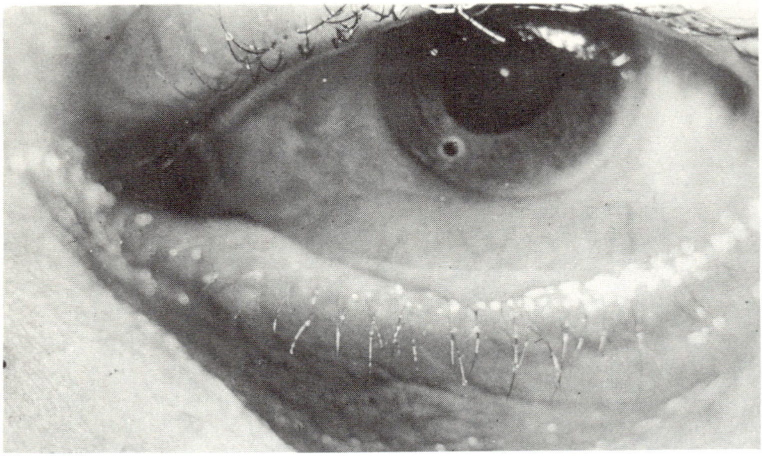
Fig 4.23

sensation, sensitivity to light, itching, burning, or a redness of the eye. Even the glands in front of the ear may become enlarged. Specific medical treatment is required for these forms of conjunctivitis.

We have all heard the saying: "Keep the room dark for the child with measles." Conjunctivitis accompanying measles often provokes severe photophobia (sensitivity to light). Although the light itself will not hurt the eyes, it does agitate an already irritable child.

Since acute conjunctivitis may resemble more serious conditions (acute iritis or acute glaucoma), the patient with an acutely inflamed eye should be examined by an ophthalmologist.

Boric acid and yellow mercuric oxide, the old remedies for conjunctivitis, have now been replaced by newer, more effective agents. A word of caution: do not use eye drops or ointments that have been prescribed previously for other infections, or for other members of the family. Also do not patch the inflamed eye, as the enclosed space is a good medium for the growth of the infecting organism.

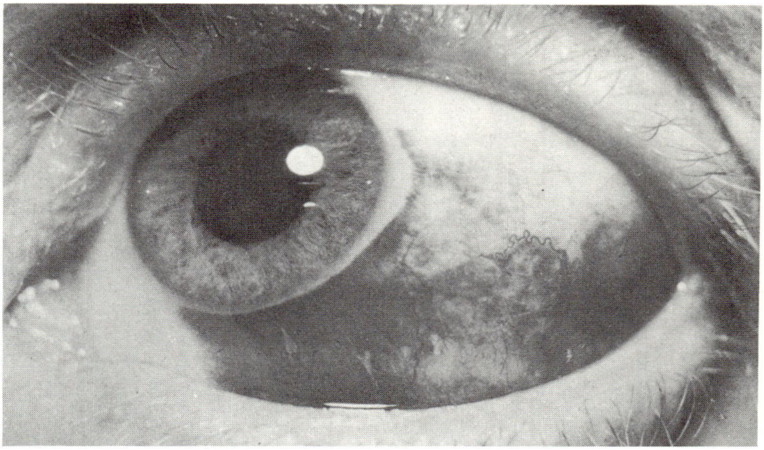

Fig 4.24

Foreign Bodies

Dust, lashes, cinders, pieces of steel, glass, or paint may fall or fly into the eye producing a painful, scratchy sensation. When a foreign body lodges in the eye, a person may believe that he can rub it out. Instead he usually embeds the foreign body in the cornea (4.23). Fortunately most foreign bodies lodge on the conjunctiva of the upper or lower lid, and they can be easily removed.

Hemorrhage

Sudden painless spontaneous bleeding on the white of the eye that does not interfere with vision is usually due to a subconjunctival hemorrhage (4.24). The appearance of this bright red blood is alarming to the patient and his family, but the bleeding is usually of no consequence because it usually absorbs in seven to ten days. It often occurs after coughing, sneezing, or vomiting severely, or after injury from rubbing the eye or being struck in the eye. Even though this subconjunctival hemorrhage may not interfere with sight, it may be an indication of serious underlying disease elsewhere in the body. Arteriosclerosis, high blood pressure, or leukemia can produce such hemorrhages.

Chemical Burns

Such chemicals as hair spray or gasoline or lye (4.25) splashed into the eye may produce serious painful conjunctivitis and dermatitis, as well as injury to the cornea (keratitis). Final visual results depend on the type and severity of the injury or burn. In the photograph of the lye burn (4.25) self-treament started before a doctor's help was sought. A hair spray keratitis may clear dramatically in 24 hours with proper treatment.

Pinguecula

This yellowish elevated fatty elastic mass of tissue in the conjunctiva is located commonly on the nasal side of the

Fig 4.25

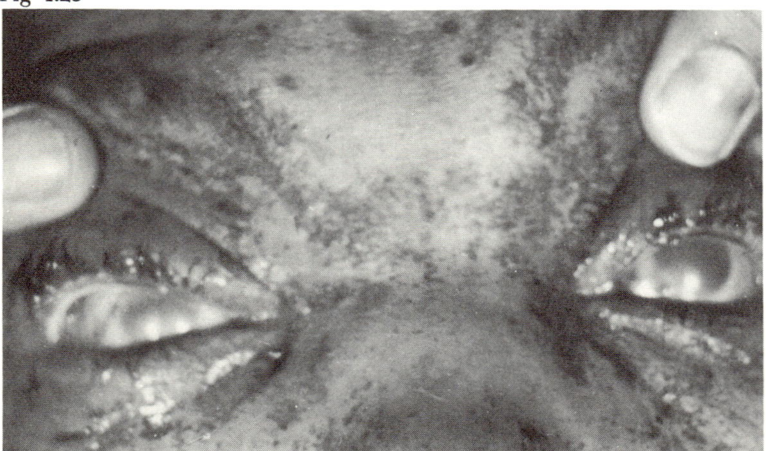

cornea (4.26). It is usually stationary and does not grow over the cornea. It may become reddened by irritation of smoke, dust, or wind causing the eye to be unpleasant looking, but it does not interfere with sight.

Discoloration

Yellowish discoloration of the conjunctiva may occur with jaundice accompanying liver disease. Bluish-gray discoloration may be the result of staining with silver salts (argyrosis), which was a common occurrence with the use of Argyrol® in treating eye or nose disorders. Brownish discoloration may be the result of diffuse nevi or increased pigmentation (melanosis bulbi).

LACRIMAL SYSTEM

The external part of the eyeball is provided with a remarkable system of irrigation and drainage—the lacrimal system. Occasionally this drainage system fails to function.

Fig **4.26**

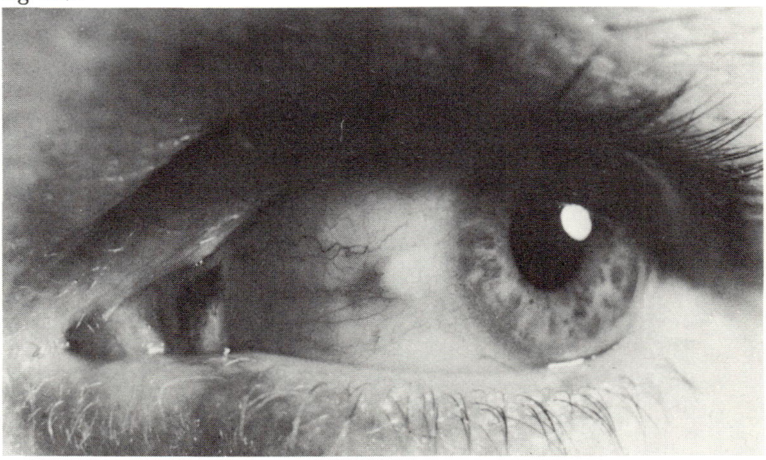

Excess Tears

In infants and adults tearing or watering of the eye may be caused by a mucous plug or concretion blocking the tear sac or duct. Since tears cannot pass into the nose, they "back up" into the eyes. If an ophthalmologist is unavailable, the following technique may be attempted without harming the eyes to help clear the condition. After washing the hands, place a finger (with a short fingernail) on the middle portion of the lower lid and gently move the finger, pressing in slightly, toward the inner corner of the eye at the nose. Then press firmly over the tear sac and, with a downward motion, bring the finger along the bone at the side of the nose. This empties the tear duct and forces the plug into the nose. Perform this procedure five or six strokes at a time, three or four times a day for a week. If there is no improvement, consult an ophthalmologist.

Tearing in the infant may also be due to abnormal development of the drainage apparatus caused by obstruction of the passage in the lid or in the duct leading into the nose. When this occurs, irrigation with a solution of fluorescein and/or probing with instruments in the tear sac and duct are necessary. Sight is not affected by this condition.

In adults tears may overflow onto the cheeks especially in cold weather. This is often due to relaxation of the skin of the lower lids, which prevents the siphoning off of tears into the punctum and tear drainage system. Overproduction of tears may also cause tearing. Irritation to the eye from smoke, chemical fumes, infection, joy, or sorrow can bring about excess tearing. Obstruction of the tear duct passageway may result from or be accompanied by a secondary infection involving the tear sac. When this occurs, acute dacryocystitis (4.27) may develop. This is characterized by swelling, redness, and tenderness over the tear sac. Pressure on the mass produces pus in the eye. Surgery is often required to alleviate these signs and symptoms and to produce a permanent cure.

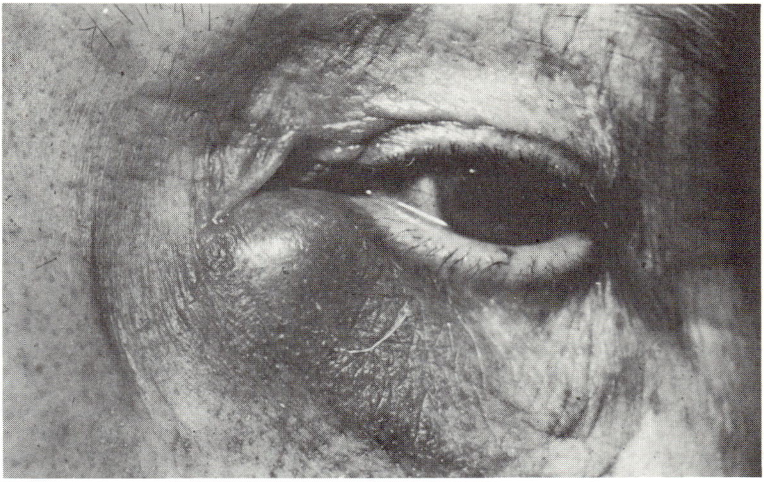

Fig 4.27

The procedure, called dacryocystorhinostomy, involves making an opening between the lacrimal sac and nose. Until an ophthalmologist can be consulted, warm compresses applied to the lesion every two hours may produce temporary relief.

Lack of Tears

This condition may occur in adults especially at the time of change in life. It may be related to an endocrine disturbance, an inflammation of the lacrimal gland itself (dacryoadenitis), a tumor involving the lacrimal gland, or paralyis of the nerve to the gland. Drying of the cornea can produce keratitis sicca, which is characterized by an irritating, scratchy sensation in the eye or just red eyes. Failure to close the lids or infrequent blinking can also produce a drying of the cornea called exposure keratitis. The symptoms of both conditions can be relieved by artificial tears and patching the eye, but the patient should be observed closely by a physician to prevent complications.

EYE MUSCLES

Strabismus (Squint)

Strabismus and squint are terms employed to indicate external eye muscle imbalance that prevents the two eyes from being straight and working together. In this context a squinting eye means an eye that deviates in or out or up and down, not an eye that appears to be contracted or winking or blinking. Since an eye is crossed only in relation to the other eye, both eyes are affected by muscular imbalance.

If a person has normal vision with both eyes (binocular vision), lines drawn from a viewed distant object to the respective center of vision in each eye are parallel. This is known as the visual axis. If the eyes deviate, the lines will not be parallel and the person may have diplopia, or double vision. A tendency for the eyes to deviate is heterophoria; actual persistent deviation is heterotropia or strabismus.

Fig 4.28

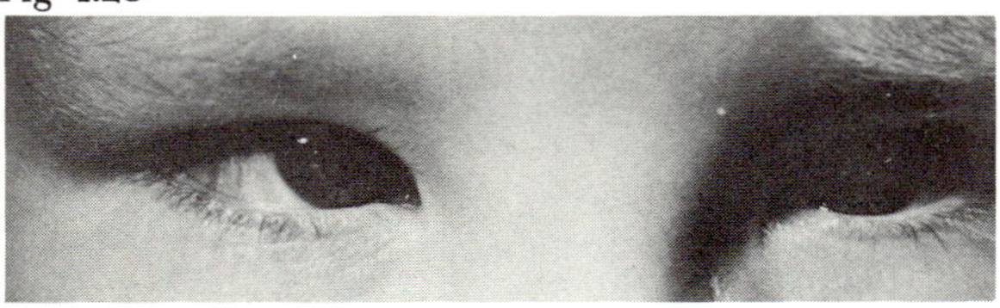

Fig 4.29

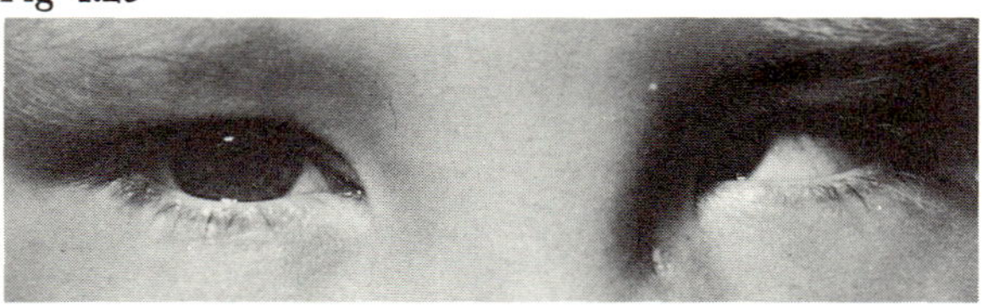

Fig 4.30

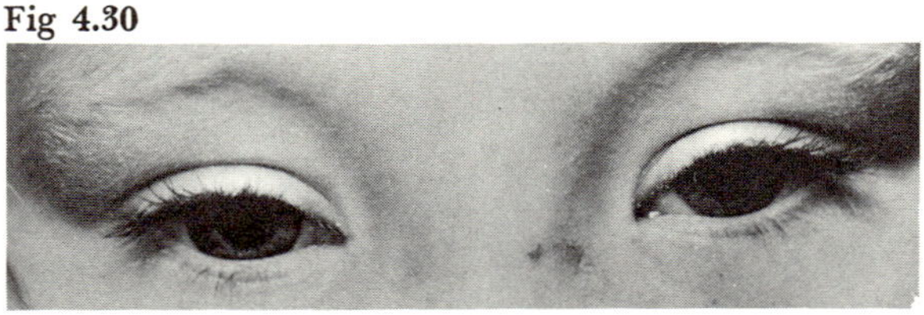

ESOTROPIA

Convergent strabismus, internal squint, or "crossed eye" exists when one eye turns in toward the nose and the other is straight. The turned in eye exhibits little whiteness, as compared to the other eye. Only one eye may be crossed, or first one eye may cross and then the other or alternator (4.28 and 4.29). Surgery straightened the eyes (4.30).

EXOTROPIA

Divergent strabismus, external squint, or "walled eye" exists when one eye turns out toward the temple and the other eye is straight. The turned out eye reveals more whiteness than the other eye. Only one eye may turn out, or first one eye may turn out and then the other (4.31 and 4.32). Again, surgery straightens the eyes (4.33). The exotropia is permanent if one eye always turns out. It is intermittent if one eye turns out during fatigue, fever, emotional crisis, illness, or under the influence of alcohol.

Fig **4.31**

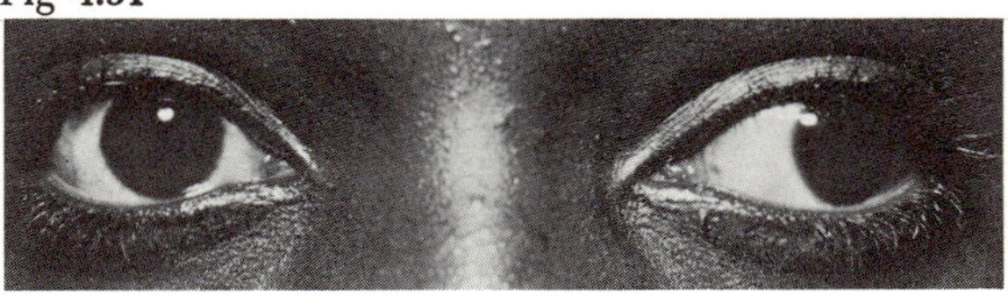

Fig **4.32**

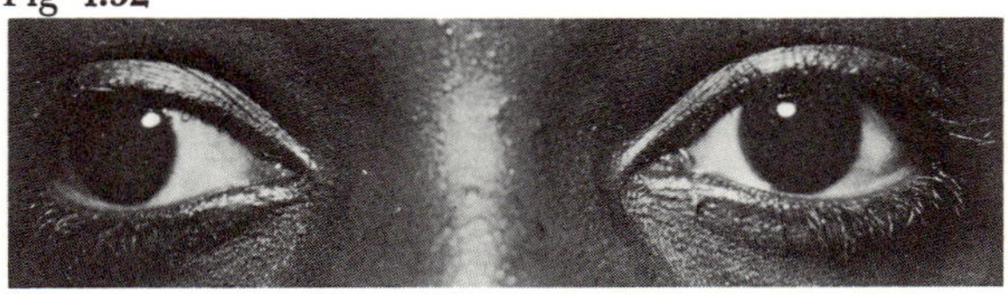

Fig **4.33**

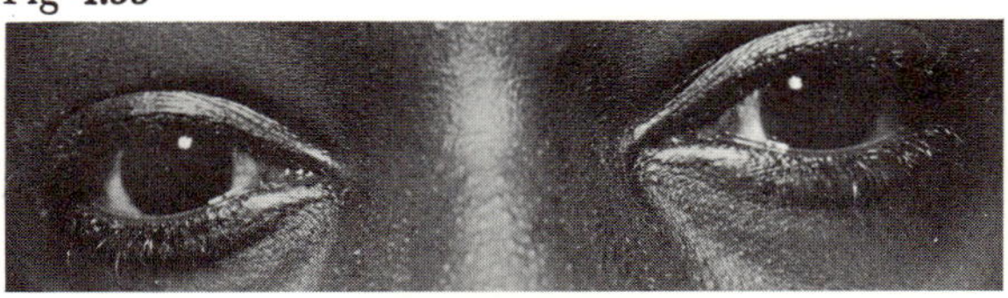

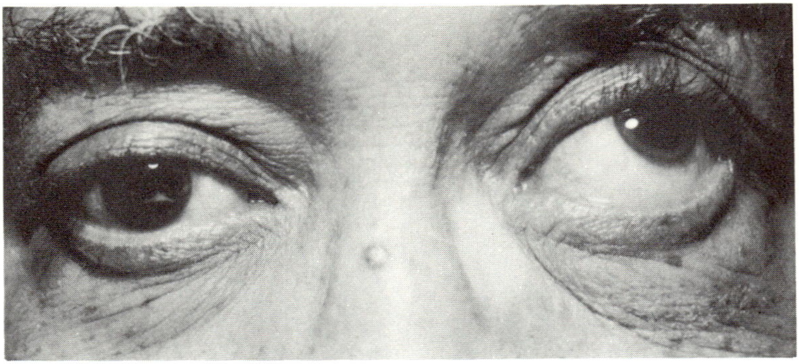

Fig 4.34

HYPERTROPIA

Hypertropia exists when one eye is elevated. This muscular imbalance may accompany either esotropia or exotropia. The white portion of the eye may be seen below the elevated eye (4.34).

WHEN SHOULD A CHILD BE EXAMINED FOR STRABISMUS?

Although it is commonly thought that a child will outgrow a crossed eye, 95 of 100 children with strabismus will not. Whether the child is 6 months or 2 years old, he should be examined as soon as his parents note the defect. The onset of strabismus is often noted when a child attempts to look closely at his picture books, after an illness such as whooping cough or measles, or after a violent fall. These occasions may be associated with the presence of the strabismus, or they may be coincidental.

Frequent blinking or squinting is not related to strabismus, but it may be induced by infection, habit, the desire to mimic a friend, or nervousness. Playful crossing of the eyes will not produce any permanent eye defect. Some exercises are even prescribed to encourage crossing the eyes toward the nose.

Prominent skin folds at the inner corner of each eye called epicanthal folds may give the false impression that the child's eyes are crossed, when they are actually straight. When the nose develops and the folds retract and disappear at about 5 to 6 years of age, the eyes appear to be straight.

When an eye is turned in (crossed) or turned out (walled), the child will experience diplopia. To prevent the confusion of double vision, the brain causes the turning eye to become lazy, or amblyopic, through suppression, or consciously or unconsciously closing off or holding back the function of that eye. *If this condition is not corrected before the age of 6 years, the child will have poor vision in that eye for the rest of his life. Any straightening of the eye after that time will merely improve the patient's physical appearance, not his vision.*

To "wait until the child is older" is to wait until it is too late. The eye will have developed amblyopia ex anopsia (poor vision resulting from disuse) from crossing or squinting. The older the child, the less the likelihood of restoring useful vision. If a child is treated by the age of 6, he should have good equal vision and straight eyes. After that time, his vision may be only partially improved. He will not have straight central vision with the poor eye while looking directly at a subject, as in reading, but he will have side vision.

When the two eyes are straight and seeing equally, the brain unites the two images into one. Depth perception, which is also called stereopsis or fusion, is the sensation enjoyed at a 3-D movie or when looking through a stereoscopic viewer. This phenomenon is well developed by the age of 6 years.

HOW IS STRABISMUS DETECTED?

If the eyes are tested by the cover test method, first one eye and then the other is covered. Any deviation of the eyes is noted as the occluder is moved from one eye to the other.

Another method of detecting muscular imbalance is by comparing the amount of whiteness that exists in the two eyes between the colored part and the inner corner of the eye.

In another test light is flashed into the patient's eyes from a flashlight held one foot away. Normally the reflex of the light striking the cornea is in the center of both pupils. If the light reflex is in the center of one pupil (the straight eye), but is off the pupillary center and on the outer temporal corner of the other (the deviating eye), the patient has an esotropia. If the light reflex is in the center of one pupil and on the inner (nasal) corner of the other eye, he has an exotropia since that eye turns out.

In still another means of determining a muscle imbalance a red lens with lines, called a Maddox rod, is put before one eye to convert the light to a line or streak. The patient is asked to look at a light held 14 inches away for near vision and 20 feet away for distant vision. Then the relationship between the line and the light is determined. If they are superimposed, no imbalance is present. If they are apart, the deviation is measured with prisms. These lenses displace the apparent position of the object to maintain the parallel visual axis of the two eyes. By using the various strengths of prismatic lenses, the doctor can measure the degree of deviation of the tropic (turning) eye from the parallel position. Both horizontal and vertical turning components are measured for distant and near vision.

During a complete eye examination of a child at the age of 3 or 4 years, a slightly crossed eye with poor vision may be detected by any of these tests.

After the child is first examined, the doctor usually tells the mother to instill atropine drops or ointment into the child's eye for three days. (Doctors have individual preferences regarding pupil dilating medication and the manner of administration. Some may prefer to administer weaker, shorter acting cycloplegic drops during a single office visit.)

Atropine is used because it is the most powerful cycloplegic drug. When the pupil is dilated and accommodation is temporarily paralyzed by atropine, all refractive errors become apparent to the doctor. One in 500 children develops a sensitivity reaction to this cycloplegic drug, and his face becomes red and flushed. The mother should not become alarmed; she should simply discontinue the drug. In the short duration in which atropine is used severe drug sensitivity is rare. If the mother gets any of the medication on her fingers and then rubs her eyes, her pupils will also dilate for a period of ten days, causing her difficulty with close work.

The mother plays the flashlight game (flashing lights in the child's eyes so that he will not be afraid of the lights in the doctor's office) and the "E" game with her child. In this game the mother cuts two "Es" out of cardboard and gives the child one, as she holds the other. She then asks the child to point the legs of the "E" (4.35) in the same direction that she points (up, down, left, and right). The child then plays

Fig 4.35

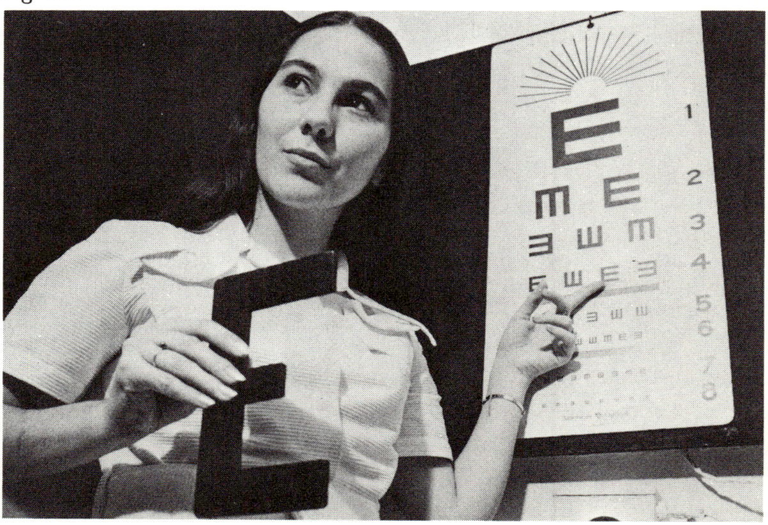

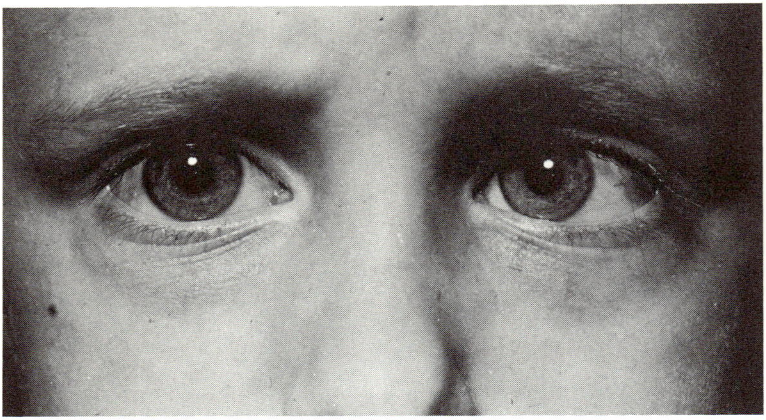

Fig 4.36

the "E" game with the "E" chart in the doctors office, and his visual acuity is determined accurately.

In a few days the child's eyes are reexamined and refracted to determine whether glasses will be needed. Through the dilated pupil, the fundus or back of the eye can be examined for tumors, scars, or other diseases that could interfere with sight and cause the eye to turn in or out. After this complete examination, a course of treatment is chosen.

HOW ARE THE EYES STRAIGHTENED?

The ultimate aims in the treatment of strabismus are:

- To obtain equal vision in each eye.
- To correct the eyes anatomically so that they appear straight (4.30 and 4.33).
- To establish fusion or stereopsis through exercises called orthoptics.

Eyes may be straightened by glasses or by the use of eye drops or ointment. The strong straight eye may be patched to strengthen the vision of the weaker deviating eye. Muscle exercises may be prescribed. Surgery may be required.

Glasses. Most children with esotropia are hyperopic or

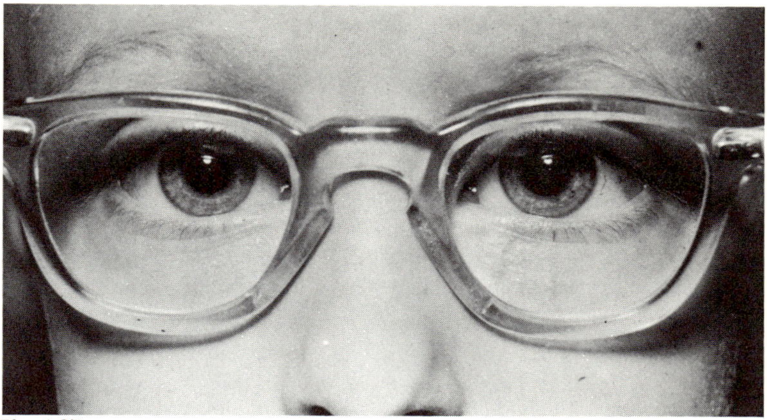

Fig **4.37**

farsighted; most of those with exotropia are myopic or near-sighted. Glasses not only help the child to see better, but also help to straighten the eye. When a child's eyes are farsighted and crossed and straightened with glasses, the condition is called **accommodative strabismus** (4.36 and 4.37). However, if the child with crossed eyes is not farsighted, it is unlikely the lenses will help to straighten his eyes.

Eye drops or **ointment** may be instilled into the child's eyes daily to constrict the pupils. This prevents or decreases the turning in of the eye in an accommodative strabismus. Medication or glasses may be used alone or in combination to straighten the eyes.

Patching or **occluding** the dominant or straight eye with elastic bandage forces the deviating lazy suppressed eye to improve its central vision. Occlusion must be constant and complete for four to six weeks to produce any results in the weak, lazy, amblyopic eye. Patching for one to two hours a day has little effect compared to total occlusion: one hour's peeking may negate one week's patching of the eye. The patched eye will not suffer any appreciable loss of vision during the patching program. If the child is very young (3 to 4 years of age), vision in the patched eye may become somewhat worse, but sight in that eye will always return to its original state when the patch is removed.

Patching is not necessarily performed if the child has **intermittent** or **alternating strabismus.** Intermittent strabismus usually occurs after fatigue, alcohol consumption, emotional excitement, or illness. At these times the eyes may diverge temporarily or turn out, or they may cross and turn in. Alternating strabismus indicates good vision in each eye; each eye fixes well separately while the fellow eye crosses or turns out. Alternators usually do not benefit from glasses. Although red nail polish may be applied to the spectacle of the dominant eye and eye exercises may be attempted, patients with alternating strabismus usually require surgical correction.

Muscle exercises. When the two eyes have equal or near equal vision and are almost anatomically straight, orthoptics or visual training is prescribed to encourage the two eyes to

Fig 4.38

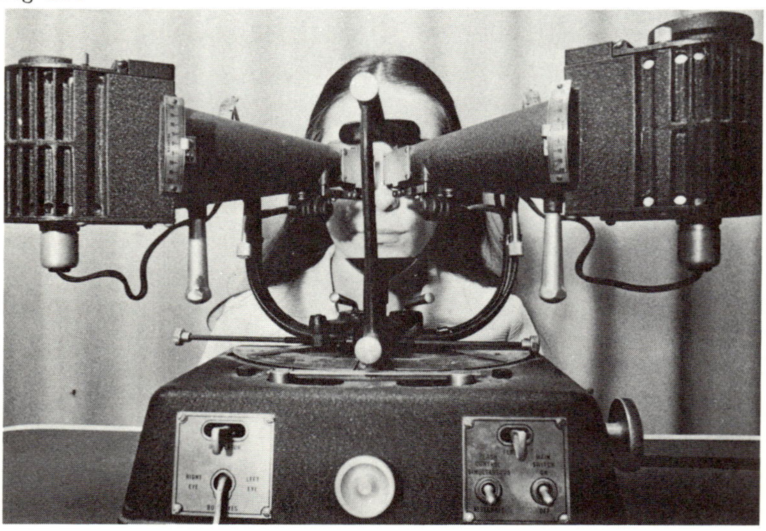

work together. Various orthoptic machines (4.38) may be used to stimulate this process and to help develop fusion or stereopsis. Other exercises carried out in the office of the doctor or orthoptic technician or at the patient's home are used for the same purpose. These exercises will help to develop some degree of depth if only a small deviation is present, as images from each eye are superimposed in the brain. If visual training or exercise does not bring improvement in six months' time, surgery is considered. Orthoptics are begun again postoperatively. If a large degree of squint is present, all the visual training in the world will not straighten the eyes and make them work together. Surgery is required, and exercise may be prescribed after this treatment.

Corrective surgery. Before surgery, parents are briefed about what surgery will be performed and usually how many muscles will be operated upon. They are also told it may take one, two, or even three operations to straighten the eyes. The surgeon cannot always predict the condition of the muscles until he operates; many variables enter into the repair of human eye muscles. However, most patients require only one operation, which is usually performed painlessly under local or general anesthesia depending on the age of the patient. The aim of surgery is to correct the position of the eyes by changing the attachments of the muscles to produce straight eyes.

The surgeon may recess (weaken or set back) the muscle that pulls too much by cutting the muscle from its site of insertion on the eyeball and reattaching or sewing it further back on the eyeball. He may resect (strengthen or shorten) the muscles that do not pull enough by taking off a portion of the muscle and reattaching it to the eyeball at or in front of its site of insertion. Catgut sutures are used in attaching the muscles to the eyeball and for closing the wound. They do not have to be removed, as they usually absorb or fall out in 10 to 15 days. The eye is not brought out on the cheek or scraped; it is rotated in the socket or orbit.

PREPARING THE CHILD FOR SURGERY. The child is told he will go to the hospital to have his eyes straightened. At the same time he is taught to play the ice cream or telephone game: he talks into the "gas machine" and tells the doctor what kind of ice cream or what telephone number he wants. He learns the peek-a-boo game in which both eyes are patched after the operation. (Some doctors, however, do not patch the eye.) He is told that he will be with other children, have a good time, and come home with straight eyes. His mother washes his hair and instills antibiotic eye drops for several days before he goes to the hospital.

HOSPITAL ROUTINE. Children frequently bring a favorite doll, toy, or blanket from home to make them feel more secure in the hospital. The child is admitted the night before surgery, and hospital gowns may be furnished. The operation usually lasts one hour, depending on the number of muscles operated upon. After surgery the child goes to the recovery room for a short time, with both eyes patched, and then to his hospital room. He is usually allowed to sit up in bed on the day of surgery, and he may get up to use the bathroom with help. The patches usually remain in place during the hospital stay of one or two days. Glasses worn before surgery are usually worn after surgery. Sun glasses are recommended after hospital discharge so that the irritable eye is as comfortable as possible.

CARE AFTER SURGERY. The mother bathes the child's eyes three times a day for two to ten minutes with cold compresses for 48 hours. After this time, warm and cold compresses are alternated to alleviate swelling and redness. Antibiotic drops or ointment is instilled into the eye or eyes that have been operated upon. Several days later the child is examined in the ophthalmologist's office. He may return to school one or two weeks after surgery. He should not go swimming for about four weeks. After a month, the eyes are well healed, the redness has subsided, and the nerves and muscles have resumed their normal function.

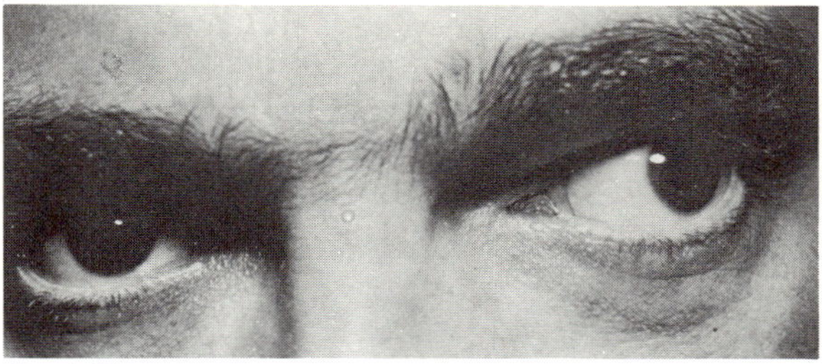

Fig 4.39

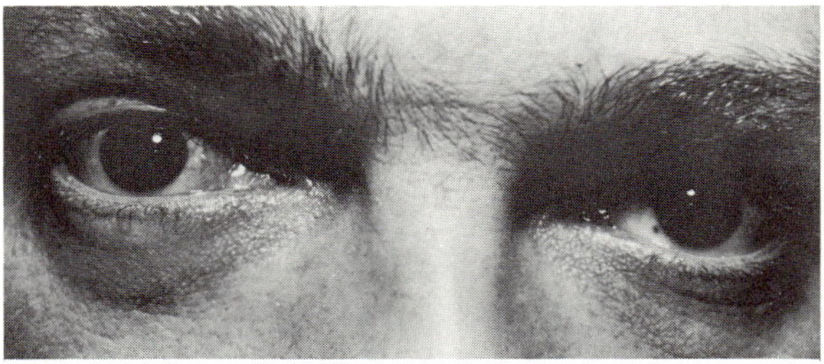

Fig 4.40

After surgery, glasses, eye drops, and orthoptic visual training may be prescribed to bring about a more permanent cure. Glasses may be used not only to help straighten the eyes, but also to maintain good vision, as is the case in partial accommodative strabismus. The operation does not change the shape of the eyeball or affect the vision or sight; it only changes the location of the muscle attachment to the eyeball. It is impossible to predict which patients will need a change of glasses after surgery.

Cosmetic surgery. Since we live in a society with other people and wish to be attractive to them, a child over age 6 or an adult suffering from esotropia or exotropia with uncorrectable vision can still enjoy cosmetically straight eyes through plastic muscle surgery. The patient will be happier and will adjust better to his handicap. His friends will not have to wonder whether he is looking at them or at someone else across a room (4.39 and 4.40).

STRABISMUS AND HEREDITY

Strabismus exhibits a dominant pattern of inheritance. In other words the condition affects the individual, his antecedents, and his brothers and sisters. A thorough investigation of the family history often provides important information concerning the diagnosis and treatment of strabismus. If one child in a family has a crossed eye or a parent has strabismus, all children in the family should be examined routinely before age 4 so that an unobserved defect might be discovered. Parents who have had their eyes straightened as children should have their children's eyes examined before age 3.

The following case report illustrates the importance of parental alertness and early examination:

> This family had no history of crossed eyes. However, a visiting grandmother thought she noticed one of the 5-year-old granddaughter's eyes turning in. Since no one in the family could verify this, the parents sought a professional opinion. While examining the child, the doctor noted a slight degree of crossed eye. More importantly the child could not fix well on a test target with that eye due to poor vision. After the good straight eye was patched, vision improved in the deviating crossed eye. The child then underwent successful muscle surgery. On her sixth birthday she had 20/20 vision in each eye, with third-degree fusion (the best type of depth perception) and straight eyes.
>
> During the course of therapy, it was recommended that all the children in the family be examined, since crossed eyes frequently occur in other members of a family. "But my son is only 3 years old and is fine, has no complaints, sees well, and has no crossed eye," protested the rather indignant mother. To the surprise of all except the ophthalmologist, the child had a condition almost identical to the one exhibited by his older sister. He underwent the same course of therapy. Since no refractive error was present, neither child required glasses before or after surgery. Had the grandmother not been an astute observer and had the parents not hastened to the ophthalmologist, both children

could have had very poor vision in one eye for the rest of their lives.

STRABISMUS IN ADULTS

Adults are less commonly affected by strabismus, but it may result from paralysis of single muscles or groups of muscles due to disorders of the nervous system or lesions in the brain. The adult patient should be seen as soon as possible after the onset of the strabismus. He usually requires a complete neurologic examination to determine the cause for the squint. If no cause is found, muscle surgery is often performed to restore normal alignment and improve cosmetic appearance (4.39 and 4.40). Surgery in adults may be performed under local or general anesthesia.

Nystagmus

Nystagmus is an involuntary wandering fluctuation of the eyes in a back and forth rhythmic oscillating movement. This disorder may or may not accompany strabismus. It is usually permanent and may be due to diseases in the retina, poor development of the macula lutea, or disease in the brain. Since the cause is variable and different types of nystagmus exist, children should be seen by their doctor as soon as the defect is observed.

CORNEA

The cornea is the front transparent window of the eye. Although it can be examined directly with a flashlight and a magnifying lens, it is best visualized with a slit lamp (4.41). The powerful illumination and the high magnification of this instrument permit the physician to study the anterior ocular segment with both of his eyes at the same time to obtain depth perception. A block or a beam of light is directed so that the layers of the cornea can be studied in detail to determine the exact depth of corneal disease. The slit lamp also permits study of the anterior chamber for white cells and the investigation of diseases of the iris, lens, and vitreous.

Megalocornea and Microcornea

Megalocornea is a developmental corneal defect characterized by a very large cornea. A similar enlargement may be observed with congenital glaucoma. Microcornea is the opposite defect, a very small cornea.

Foreign Bodies

Since the abundant nerve supply of the cornea makes it one of the most sensitive parts of the body, it serves as an excellent "watchdog" for foreign material entering the eye. Dirt or specks lodging in the eye may produce scratching, knife-cutting sensations that the sensitive corneal nerves trans-

Fig 4.41

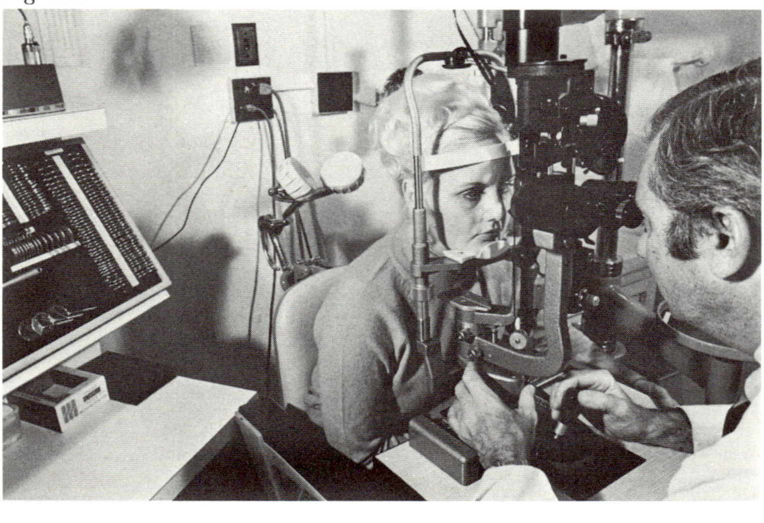

mit to the brain. If the cornea loses this sensitivity due to injury or impairment by disease, it loses its protective function. Foreign bodies may embed in the cornea (4.23).

Injuries

A twig of a tree, a piece of paper, or a fingernail as well as burns from acid or alkalies can produce corneal abrasions. If the cornea is exposed to secondary infection by a virus or bacteria, a corneal ulcer may develop (4.42). An ulcer usually will not clear by itself; it can be very serious, requiring immediate and drastic medical and/or surgical treatment. After healing takes place, vision may be impaired due to corneal opacities or scars, even though the rest of the eyeball is normal.

Corneal contact lenses also can produce a scratchy sensation and an irritable eye from a corneal abrasion. Until a physician can be consulted, the lens should be removed and the eye patched.

Fig 4.42

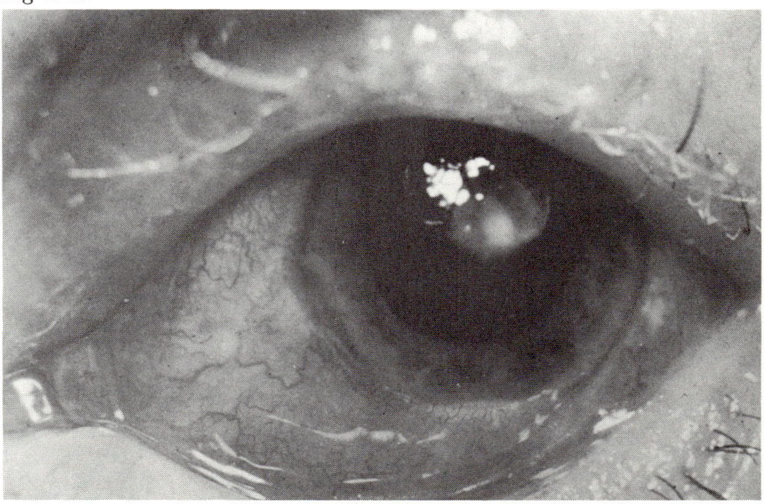

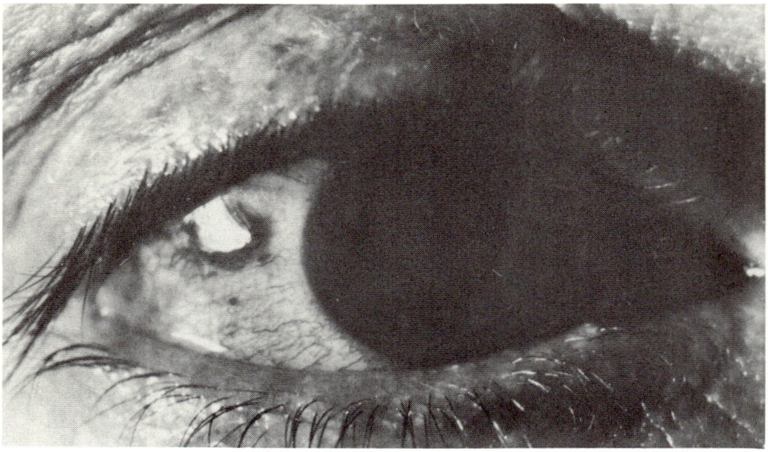

Fig 4.43

Blood Staining

This condition may result from hemorrhage in the anterior chamber (hyphema) after injury or surgery to the eye (4.43). Increased intraocular pressure usually has been present for some time, and blood has been forced into the cornea. Although the cornea may have a brownish-gray cast, this discoloration may disappear gradually with treatment and time.

Infections

Inflammation of the cornea, or keratitis, may be secondary to conjunctivitis, blepharitis, or injury. Keratitis is characterized by a painful red eye, sensitivity to light, and an occasional scratching sensation upon blinking. An ulcer may develop in the cornea after its outer layer is invaded by a viral, fungal, or other infectious organism. Herpes simplex, a virus that invades the cornea after injury, produces a dendritic (branch-like) keratitis (4.44a). Herpes zoster, another viral agent, produces inflammation of the cornea, especially if

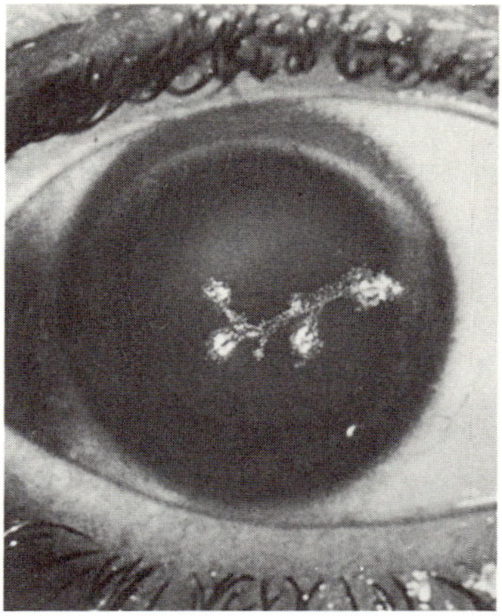

Fig 4.44a

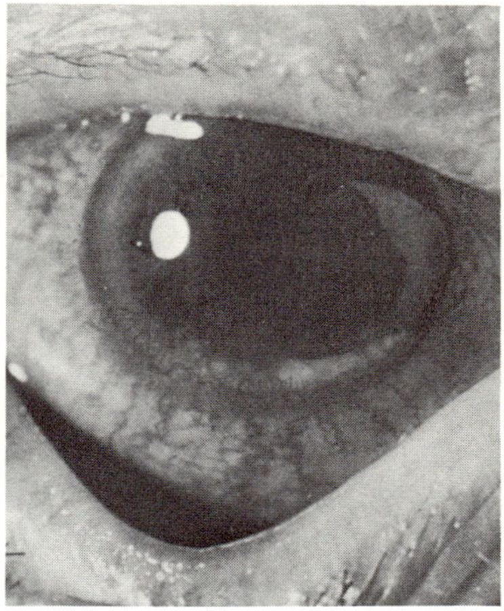

Fig 4.44b

the skin of the nose is involved. A marginal ulcer is a corneal infection that occurs near the outer edge of the cornea (4.44b). Central corneal ulcers due to infectious organisms such as pneumococcus, staphylococcus, pseudomonas, or fungi can be severe and serious; they may even cause loss of the eye. With these severe ulcers, the eye sets up a defense reaction to help fight the infection. Large numbers of white cells appear in the aqueous fluid of the anterior chamber. When these cells precipitate, they accumulate into a whitish mass called a hypopyon (4.45). This disease requires the immediate attention of an ophthalmologist.

Burns

THERMAL OR FLASH BURNS

Overexposure to intense ultraviolet rays from a sunlamp or a welding arc or torch as well as reflection from snow may produce inflammation of the eye, conjunctiva, and/or cornea called photo-ophthalmia. The eyes may become severely painful and red with tearing and light sensitivity four to eight hours after exposure to the ultraviolet light. (See emergency care pages 169 and 170.)

CHEMICAL BURNS

Acid or alkaline solutions splashed into the eye may produce a similar clinical picture. However, symptoms and signs occur immediately after exposure to the chemical and may be more severe in nature. (See emergency care page 169.)

Pterygium

This grayish elevated growth of elastic and connective tissue containing blood vessels invades and grows over the cornea (4.46). It may result from irritation to the eye from wind, heat of the sun, dust, or smoke; it is common in people from Southern climates. If the pterygium progresses to grow over the center of the cornea, sight may be impaired or even lost. Before this occurs, the pterygium should be removed surgically. Some people confuse a cataract with a

Fig 4.45

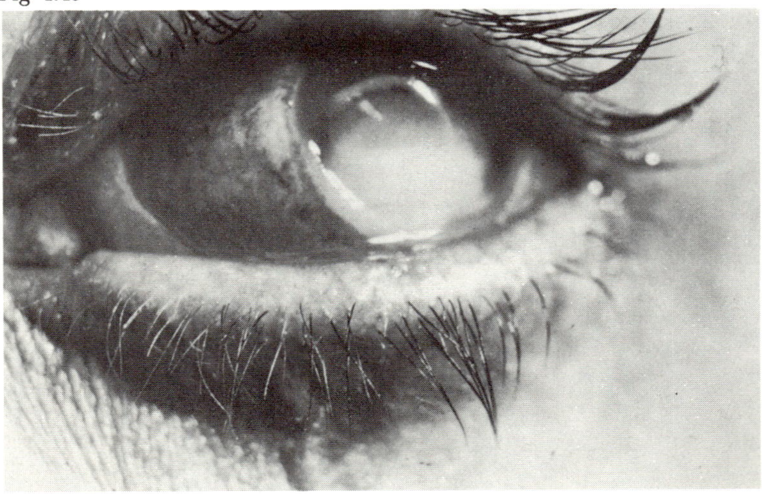

pterygium by calling a cataract a "skim or skin growing over the eye." A cataract, however, is a clouding of the lens, which is located inside the eyeball.

Degenerative or Aging Changes

Dystrophies or degenerative aging processes (4.47) may develop in the cornea and interfere with vision. They are slowly progressive, noninflammatory, and usually affect or involve both eyes. They may produce a haziness or cloudiness of the cornea. If the vision is markedly impaired, contact lenses may be prescribed to improve vision. If they do not help, a corneal transplantation may be performed to restore useful sight.

KERATOCONUS

This bulging and thinning of the cornea gives it a cone-shaped appearance. Usually involving both eyes and causing blurring of the vision, the condition produces a high degree

Fig **4.46**

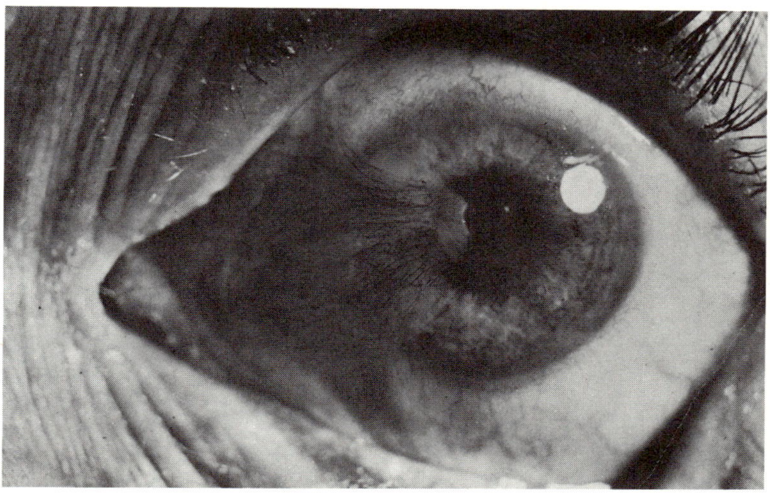

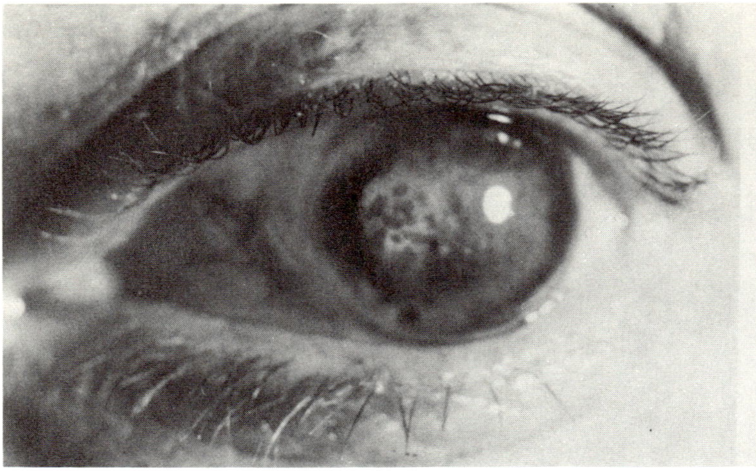

Fig 4.47

of astigmatism. This visual defect may not be corrected by spectacles; it requires the use of contact lenses or even surgery, in the form of a corneal transplantation, to improve sight.

ARCUS SENILIS

This common corneal aging process is characterized by a whitish, grayish, or yellowish colored ring located near the outside edge of the cornea (4.48). (This patient also has a cortical cataract and a pterygium.) It usually occurs in the elderly in both eyes and does not interfere with vision. In a young person this condition may indicate a high level of cholesterol in the blood.

Corneal Transplantation

When the cornea is involved by degenerative change, infection, or injury, scar tissue may form as healing occurs. If the scar involves the center of the cornea or the entire cornea, vision is impaired. Depending upon the degree of involvement, the person may not be able to see to perform his daily tasks. Contact lenses rather than spectacles may par-

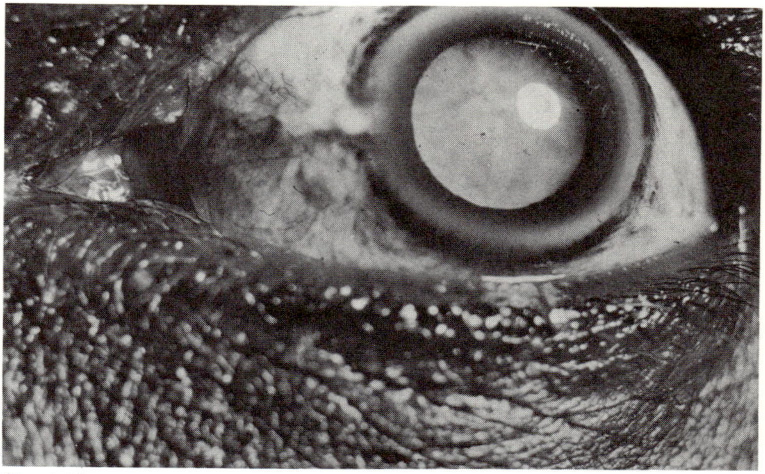

Fig 4.48

tially improve vision, but often they are ineffective and a corneal transplant is required.

WHY IS A CORNEA TRANSPLANTED?

Corneal transplantation, or keratoplasty, is an operation designed to correct partial blindness resulting from corneal disease. Eye tissue from one person is transplanted into the eye of another person who has been partially blinded by a corneal scar or disease.

Many people are under the false impression that one good eyeball is transplanted for another eyeball which is diseased. Some mistakenly believe that a blue-eyed person's eyes cannot be used for transplantation in a brown-eyed person. Neither of these statements is true. The only tissue used in the transplant is the cornea, which has nothing to do with the colored part of the eye. Since the eye is connected to the brain by the optic nerve, which is a part of the central nervous system, the eye is not and cannot be transplanted (1.3).

If the eye is compared to a watch, the crystal of the watch would be synonymous with the cornea of the eye. The face of the watch would be equivalent to the iris and lens (4.49). If the watch crystal is clean and transparent, the face of the watch will be seen clearly. However, if paint is smeared over the crystal of the watch, the face of the watch will not be seen and the paint cannot be wiped off. To see the watch face clearly again, the crystal must be removed and replaced with a new clean crystal.

HOW IS A CORNEA TRANSPLANTED?

A corneal transplantation, like a cataract operation, usually is performed under local anesthesia. General anesthesia is used for children and apprehensive or nervous patients. The operation is completely painless and takes about one hour to perform. The diseased, cloudy, opaque cornea (4.50)

Fig 4.49

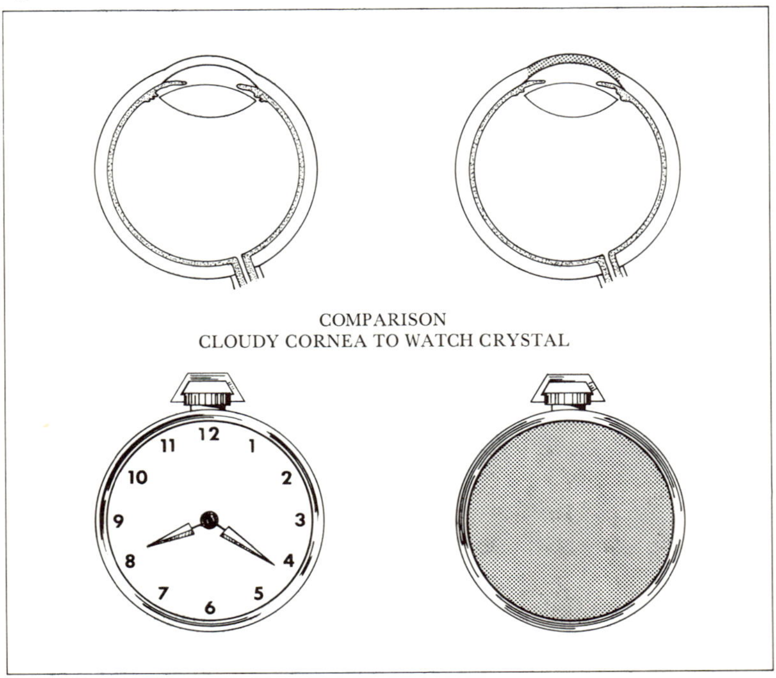

is removed from the recipient's (living patient's) eye, and replaced by a new clear cornea (graft) from the donor's (deceased person's) eye. The new cornea is then sutured or stitched into place (4.51). As few as eight and as many as 20 or more sutures may be used, according to the size of the graft, to hold the border of the graft to the border of the recipient. If the operation is successful and the graft "takes" and remains clear (4.52), the patient should see well again if the lens and the retina behind the cloudy cornea are normal. The patient is usually hospitalized from seven to ten days; he can return to work in four to six weeks.

HOW SUCCESSFUL IS CORNEAL TRANSPLANTATION?

In favorable subjects the rate of success of corneal trans-

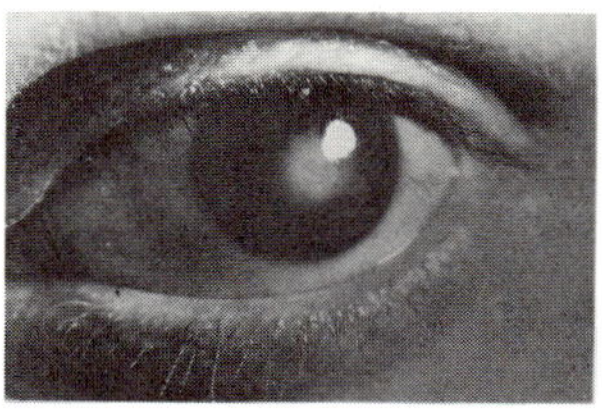

Fig 4.50

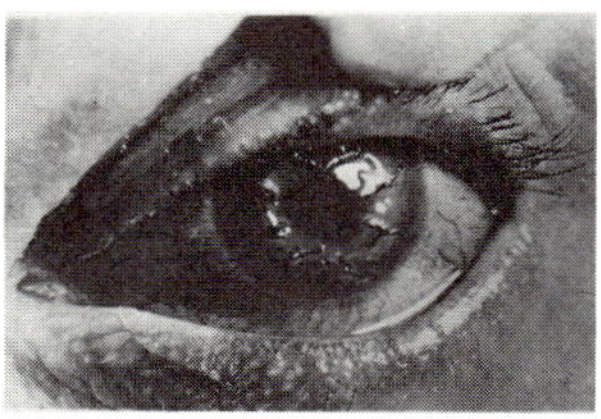

Fig 4.51

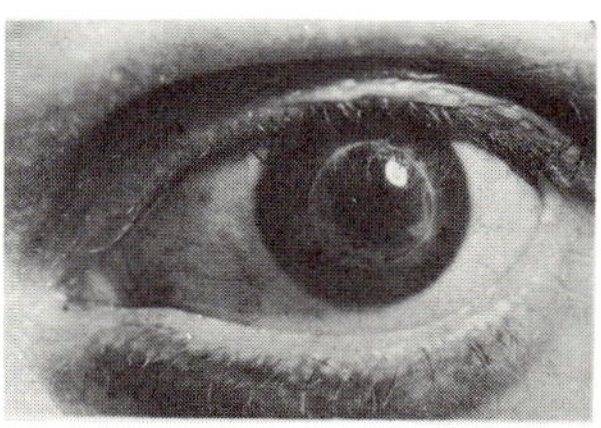

Fig 4.52

plantation may be as high as 60 to 80% with final visual acuity of 20/20 with glasses. In unfavorable subjects the rate of success may be around 15 to 20%. Each patient is evaluated individually before definite results can be predicted. The most important factors in determining the final results are:

- Basic corneal disease (some types of corneal disease respond better to corneal transplantation than others).
- State of the donor's cornea.
- Surgical technique and skill.
- Healing ability of the recipient cornea.
- Sensitivity reactions between donor and recipient cornea.

A corneal transplantation will not help every blind person to see again. If a person is blinded by glaucoma, a detached retina, or degenerative change and the retina has been damaged or destroyed, nothing can restore lost sight. Corneal transplantation restores vision only in eyes that have been partially blinded by *corneal* disease. Some vision must be present before transplantation is even contemplated.

This case report of an unusual corneal transplant illustrates the sometimes remarkable success of the operation.

> A 51-year-old man injured his right eye while working. After the accident, he developed a corneal ulcer which spread so rapidly that it could not be controlled by any form of therapy. When it began to penetrate the entire cornea, the man was advised to have his eye removed. Unhappy with this advice, he sought another doctor's opinion. When the patient was seen by us, he was told that he had a chance of having his eye saved by a therapeutic corneal transplant and, if this operation did not succeed, he could always have the eye removed.
>
> The parents of a child who had been the victim of an automobile accident willed their child's eyes to the eye bank, enabling the corneal transplant operation to be performed. The corneal ulcer was removed from the man's eye, and the healthy cornea from the child's eye was inserted

into its place. The operation was successful and the graft held. Although the cornea did not remain perfectly clear, the man obtained 20/50 vision in that eye and was able to return to work with two seeing eyes. The operation preserved this man's eye and his sight, and it saved his company from a large industrial compensation claim.

EYE BANKS

There are now many eye banks throughout the United States; the headquarters is the Eye Bank for Sight Restoration in New York City. The eye bank unlike a blood bank is not a place where eyes are stored in large quantities over a long period of time. It serves instead as a receivership for eyes obtained from deceased persons. These donor eyes must be removed under sterile conditions, using sterile technique, within two hours after death. After the eyes are removed, they are placed in a sterile glass container and sent to the pathology laboratory. Here they are processed and refrigerated at a temperature of 2 to 3°C for as long as 24 hours. At this time they must be used for corneal transplantation. After this time, the cells of the cornea, which have remained alive even though the patient has died, may be no longer viable. Corneas are used for transplants as soon as possible since this increases the chances for a "take" of the graft. If corneas are not used for transplants, the vitreous of the eye is salvaged for use in retinal detachment surgery or for research. Frozen corneas may be used for lamellar or partial thickness corneal grafts.

In recent years the use of freezing or chemicals to preserve corneas has made available more corneas for lamellar corneal transplantation. This involves removing the damaged portion of a layer of the cornea and replacing it with a new layer from the donor's cornea. Although the transplantation of animal corneas into humans has not been successful, it is still under investigation. Plastic artificial corneas (keratoprostheses) also have been substituted for human corneas. Although these prostheses (4.53) have been successfully implanted in animal

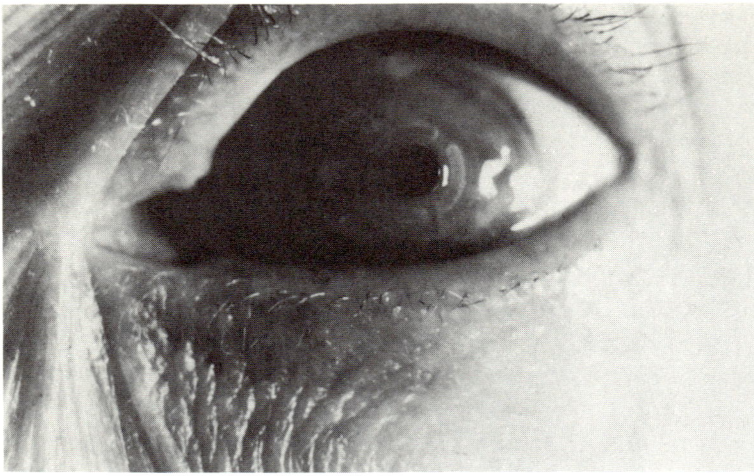

Fig 4.53

eyes and a few small series of human eyes, the artificial cornea has not yet proven to be a good substitute for the real cornea. In most instances this procedure has been carried out only in patients who have failed to respond to corneal transplantation.

Donating eyes. It is preferable to obtain a signed slip for eyes to be willed when a person is near death since it is often difficult to locate the nearest relatives and remove the eyes of the deceased within the two-hour period. Donating the eyes in a will is quite futile; wills may not be opened for many days after a person has died.

In an attempt to solve these problems members of some organizations pledge to donate their eyes so that they may be available for corneal transplantation after death. Removing the willed eye after death does not result in disfigurement. Conformers or artificial eyes are often inserted in place of the eye to preserve the normal physical appearance of the donors. It must also be stated with emphasis that eyes are never bought or sold; the thought of such a transaction would be repugnant to any reputable surgeon. Eye banks have a list of ophthalmic surgeons throughout the country who are in need of eyes. An eye removed in San Francisco may be used by an ophthalmologist in Cincinnati or Boston. The following steps should be taken to will eyes:

- Call the eye bank in your city to obtain eye bank slips. Fill them out and return them to the eye bank.
- Notify your lawyer that you have willed your eyes.
- Tell your family or nearest of kin that you have willed your eyes.
- Carry your eye bank slip with you or have it readily available in case of emergency.
- If any member of your family dies suddenly, call the eye bank. The necessary measures will be taken to obtain the desperately needed eyes.

As both laymen and members of the medical profession grow to understand corneal transplantation, more eyes may be willed. If so, the large number of persons who have been blinded by corneal disease will be the deserving inheritors of new sight.

SCLERA

The white protective coat of the eye may be involved by injury, inflammation, or aging processes.

Injuries

Injury from blunt or sharp instruments or objects can tear or cut the sclera. A blunt injury may produce a subconjunctival hemorrhage. This blood above the sclera and below the conjunctiva is bright red when fresh, but it becomes orange colored in the process of healing. Injury from a blunt object may also disturb the structures within the eye, causing, for example, a detached retina. Sharp objects may perforate the sclera, exposing the choroid, and causing a possible loss of vitreous. Although this type of injury can be serious, proper surgical treatment can save the eye. This fact is well illustrated by the following brief case report:

> A mechanic was struck while wearing his glasses by a piece of flying steel. The broken glass cut his sclera exposing the structures of the inner eye. Although it appeared that the eye would be lost, the eye was successfully operated upon, the wound closed, and the detached retina repaired. The patient regained 20/20 vision.

Inflammation

Scleritis means inflammation of the sclera, while episcleritis (4.54) denotes inflammation of the superficial or outer layers of the sclera (the space between the sclera and the conjunctiva). (This patient also has a xanthelasma on his upper lid). The eye is red and painful, sensitive to light, and tender to touch. The inflammation may be secondary to a systemic disease, such as rheumatoid arthritis, or to some infectious process in the body. It usually responds well to topical and systemic medication.

Fig 4.54

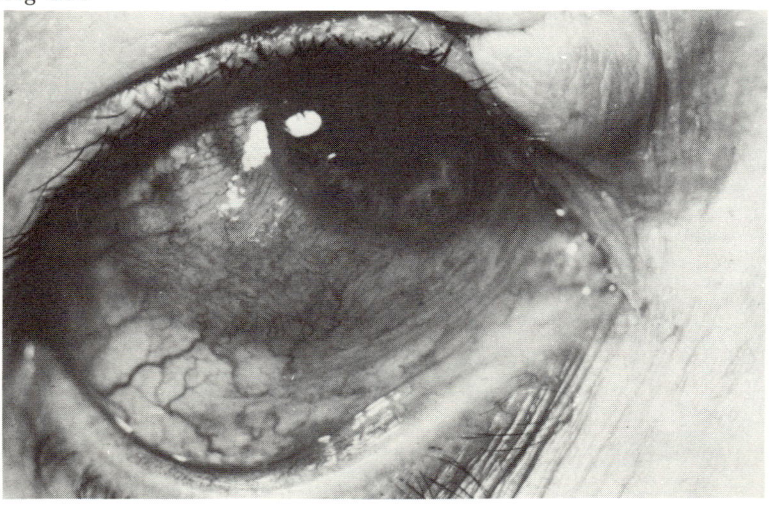

Degenerative or Aging Changes

SENILE HYALINE PLAQUE

In people over 60 years of age a gray-blue round area (4.55) develops as the sclera becomes thinned. This plaque, which may be located near the insertion of the rectus muscles to the eyeball, produces no pain or discomfort and requires no treatment. The plaque has been found to consist of calcium sulfate crystals.

SCLEROMALACIA

In this condition the sclera becomes so thin that the bluish uveal tissue is visible (4.56). Scleromalacia may accompany rheumatoid arthritis or follow severe anterior uveitis and episcleritis.

STAPHYLOMA

This disease is characterized by thinning of the sclera as well as by a bluish-purple mass that bulges outward with

Fig 4.55

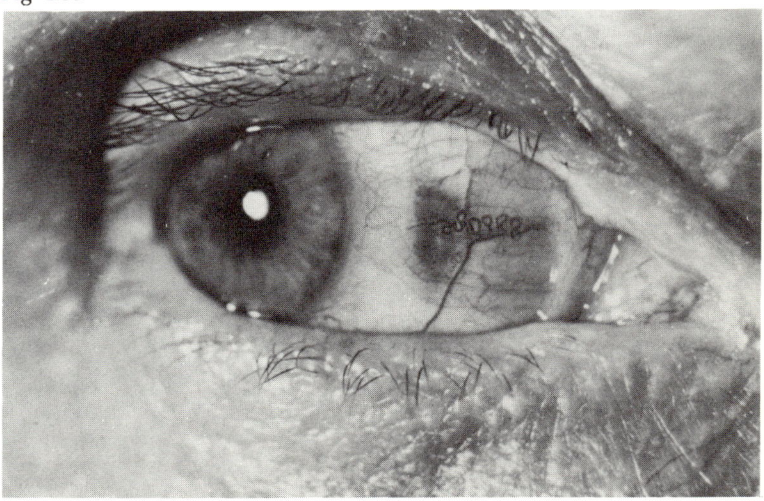

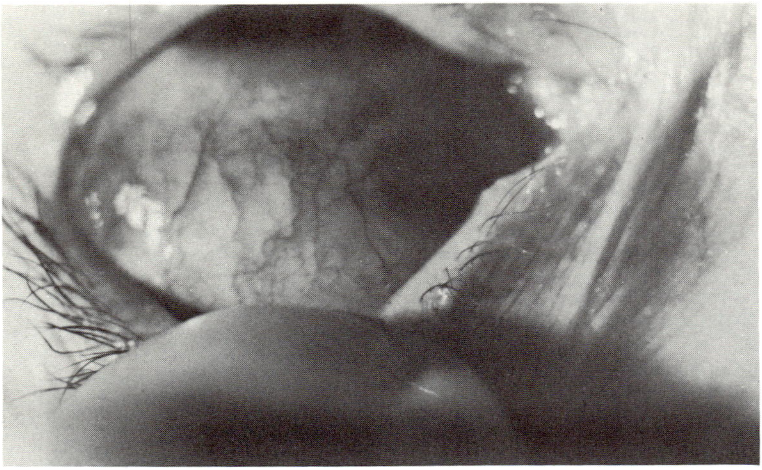

Fig 4.56

uveal tissue (iris, ciliary body, and/or choroid) under its surface (4.57). The disorder may be secondary to increased pressure in the eye from glaucoma, injury, or chronic inflammation that has thinned the scleral wall. A staphyloma of the eye is like a weak spot in a tire that bulges outward. It may be single or multiple, and may or may not produce symptoms; however, it should be observed periodically by an ophthalmologist.

UVEAL TRACT

The iris, ciliary body, and choroid together make up the uveal tract. The most common involvement of this structure is an inflammation called uveitis. Anterior uveitis involves the iris and ciliary body, which are in the front segment of the eye; posterior uveitis indicates involvement of the choroid, which is located in the back segment of the eye. Symptoms, signs, and treatment of uveitis are the same as those discussed under iritis, iridocyclitis, and chorioretinitis.

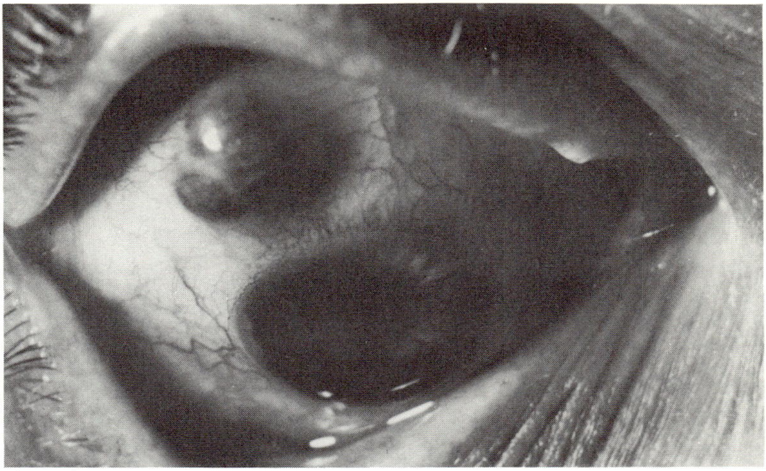

Fig 4.57

IRIS

The pigmented iris gives color to the eyes. A person with heterochromia may have one brown and one blue iris, or part of his iris may be blue and the other part may be brown (4.58). This abnormality, which may be due to heredity or to inflammation, has no effect on vision. Absence of a portion of the iris, or coloboma (4.59) may be present from birth or may be the result of surgery.

Iritis

Inflammation of the iris is called iritis (4.60). It may follow local injury to the eye or accompany an ulcer or foreign body on the cornea; it also may be secondary to such systemic diseases as tuberculosis, syphilis, sarcoidosis, or leprosy. Focal infections (badly infected teeth, tonsils, sinuses, or kidneys or diseased prostate, gallbladder, or ovaries) may occasionally cause a toxic reaction in the iris. A complete examination by an internist, a dentist, and an ear, nose, and throat specialist as well as laboratory studies may reveal the causative underlying disease process.

Symptoms of iritis include a red, painful, and light sensitive eye. Since iritis is often confused with conjunctivitis or glaucoma, the patient should be examined and treated by his physician as soon as possible. Signs of inflammation may be seen in the anterior chamber by slit lamp examination. The anterior chamber may be cloudy with white cells and flare. Flare is due to the precipitation of a protein-like exudate in the anterior chamber. It resembles dust floating in rays of sunlight coming through a window. Cells adhering to the back surface of the cornea are called KPs, or keratic precipitates. Large accumulations of white cells in the anterior chamber form a whitish mass called hypopyon (4.45). Secondary inflammation of the retina and choroid, complicated cataracts, or glaucoma could result from iritis. As healing takes place, adhesions of the iris to the capsule of the lens (posterior synechiae) or adhesions of the iris to the cornea (anterior synechiae) may develop.

Fig **4.58**

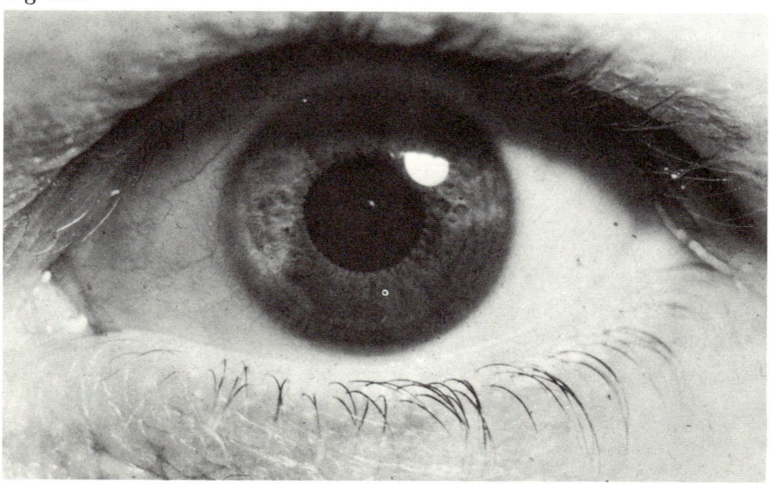

Treatment of iritis is directed toward finding and removing the causative agent or condition. Drops or ointment are instilled into the eye to dilate the pupil, put the eye at rest, relieve pain, and prevent adhesions. Steroids and antibiotics may also be used topically and systemically in combination with local applications of heat.

Tumors

Small brown areas on the iris called nevi or freckles (4.61) may be insignificant and harmless birthmarks. Large growing pigmented masses should be evaluated by an ophthalmologist to rule out the possibility of a tumor called melanoma. If this diagnosis is strongly suspected, the lesion is removed and studied microscopically.

The Iris and Glaucoma

Since the iris forms part of the drainage angle in the eye and is structurally concerned with the circulation of aqueous

Fig 4.59

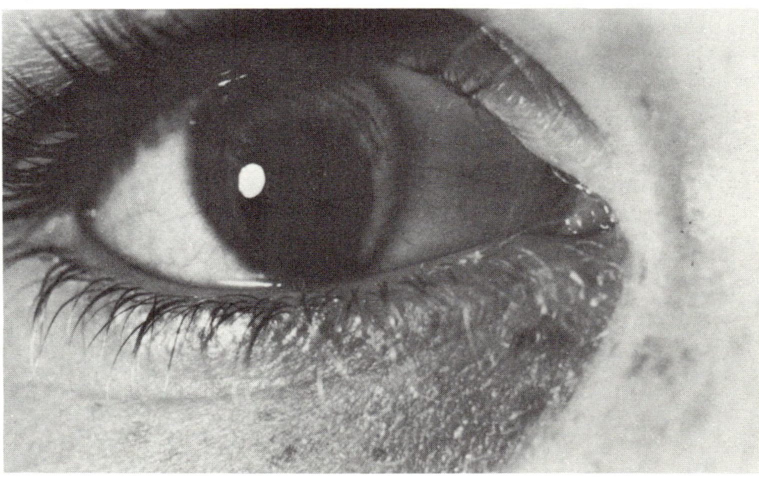

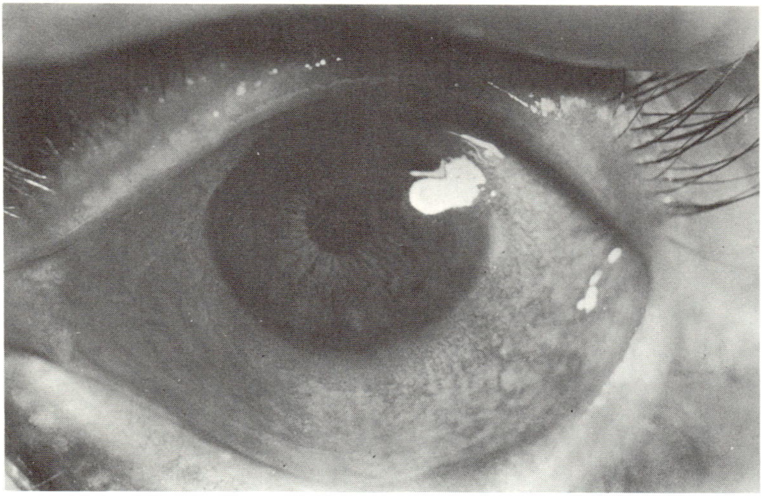

Fig 4.60

humor in the eye, it is an important structure in glaucoma. If the iris lies too close to the cornea, as in narrow-angle glaucoma, the aqueous humor may not have access to the drainage channel (Schlemm's canal), and intraocular pressure will be increased. To cure or prevent this, an opening is created surgically by removing a portion of the iris. This operation is called an iridectomy. Since an iridectomy creates permanent access to the drainage canal, the risk of a rise in intraocular pressure is avoided. Even a dilated pupil cannot block the aqueous humor from leaving the eye.

PUPIL

The pupil is the black round opening in the center of the iris. The pupils of both eyes are usually equal in size. By the action of the dilator and constrictor muscles in the iris, the size of the pupil enlarges or decreases. Nearsighted persons commonly have larger than normal pupils. The pupils

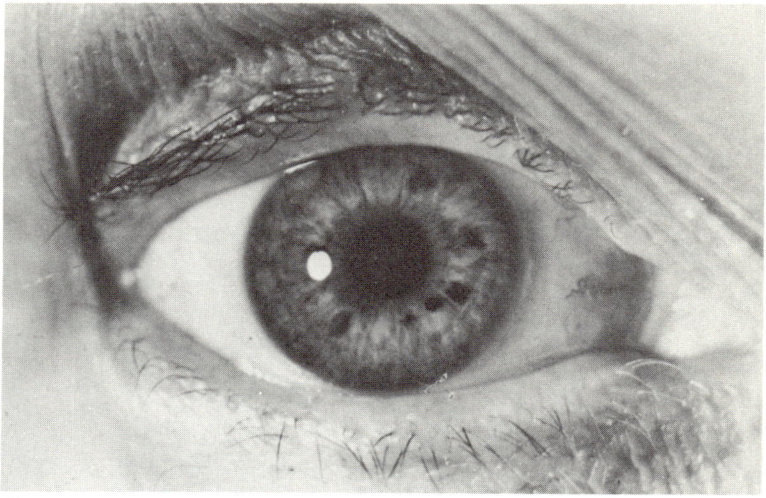

Fig 4.61

also may be dilated by the installation of cycloplegic eye drops (atropine or other belladonna derivatives) or mydriatic agents (Paredrine® or Neo-Synephrine®). Medication containing belladonna derivatives, taken internally for such conditions as gastric ulcers, may cause the pupils to dilate and produce the annoying disturbance of blurred vision for close work.

When a light is flashed into one eye, both pupils normally constrict. When the eye focuses first on a distant object and then quickly on a close object, the pupils also constrict. If the pupils do not constrict to light or constrict unequally, an ophthalmologist should be consulted. Constricted pupils may result from the use of miotics (4.62). These topical eye drops, which narrow the pupil by pulling the iris away from its root, are used in the treatment of glaucoma. Pilocarpine and phospholine iodide are commonly used miotics. Narcotics, such as morphine or heroin, will also constrict the pupils. If one pupil is more dilated than the other and will not constrict with light, there may be some disorder in the nervous system. Syphilis, meningitis, hemorrhage, aneurysm, tumor of the brain, or even severe retinal disease can produce such a disorder. Any pupillary disturbance, therefore, indicates the need for a complete eye examination.

CILIARY BODY

This muscular structure cannot be seen by the naked eye. Located at the root of the iris, its front portion lies behind the angle of the eye and its back portion joins with the choroid.

Cyclitis

Cyclitis, or inflammation of the ciliary body, is usually accompanied by iritis. Inflammation of the ciliary body and the iris at the same time is called iridocyclitis. Since the condition is clinically similar to iritis, cyclitis is treated in the same manner.

Fig 4.62

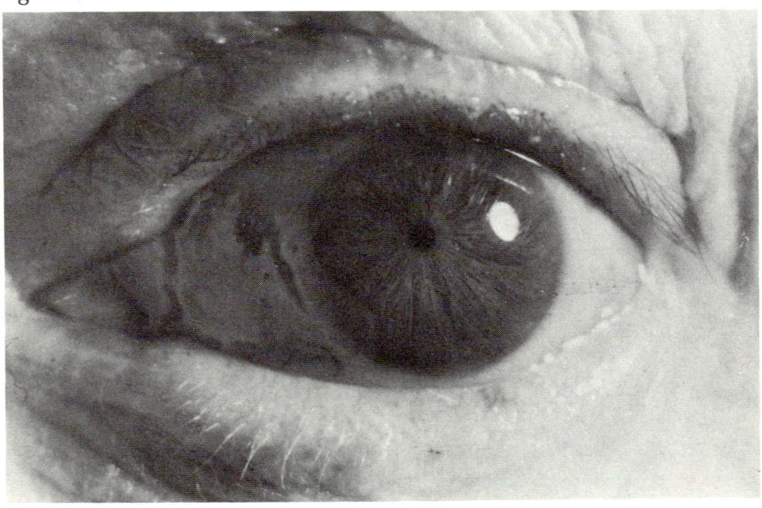

Tumors

Although tumors of the ciliary body are rare, they are usually malignant and, commonly, melanomas. A small tumor may be removed surgically by a cyclectomy; a large tumor may necessitate removal of the eye (enucleation) even if vision is good to save the patient's life. This short case report illustrates a favorable outcome of enucleation:

> When a man complaining of slight blurring of his sight was examined, a mass was discovered inside one eye. Further studies indicated that this mass was a large malignant melanoma. Even though the patient's vision was 20/30, it was recommended that his eye be removed. Pathologic reports after enucleation corroborated the clinical diagnosis. Today, ten years later, the man is alive and well.

Sympathetic Ophthalmia

Severe injury to the ciliary body of one eye may, after months or years, cause uveitis in the uninvolved eye. Since severe visual loss may occur in both eyes, a severely damaged eye is removed soon after injury, or treatment with steroids is begun as soon as signs of inflammation appear in the good eye.

CHOROID

This vascular layer of the eye can be seen only by the naked eye when the sclera has been thinned due to uncontrolled glaucoma or a chronic inflammation such as uveitis. When the eye bulges in a weak spot, the choroid is visible as a bluish-purple mass under the sclera. This mass is called a staphyloma (4.57). Because of its many blood vessels, the choroid may be involved by injury, inflammation, or tumor.

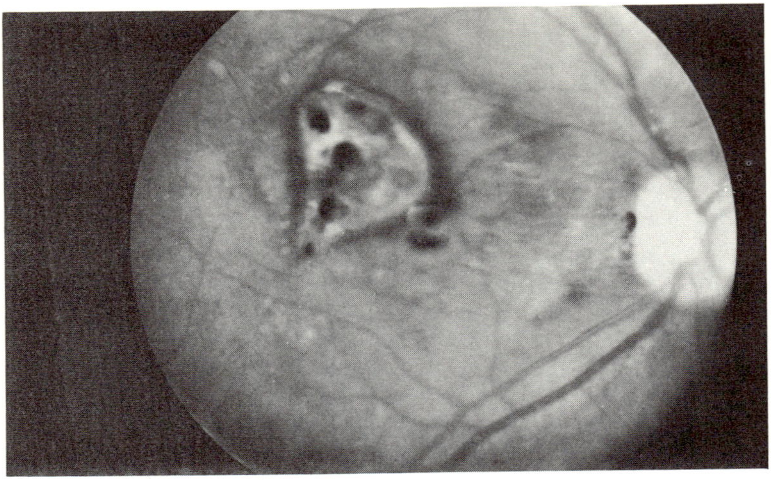

Fig 4.63

Injuries

Injuries to the eye by a blunt object may produce a choroidal hemorrhage. Since this could interfere with sight, immediate medical treatment is required.

Choroiditis and Chorioretinitis

Inflammation of the choroid is called choroiditis. Because the choroid is close to the retina, inflammation usually involves both structures (chorioretinitis). Blurry or foggy vision is a common symptom. Since choroiditis, like iritis, may be secondary to infection elsewhere in the body, a complete medical exam should precede topical or systemic treatment.

Central chorioretinitis (4.63) involves the central portion of sight, the macula lutea, causing impairment of central straight vision. If a scar forms in the healing process, central sight may be permanently impaired. Although the person will not be blind, he will have difficulty seeing objects in his direct path of vision. Powerful magnifying glasses or telescopic

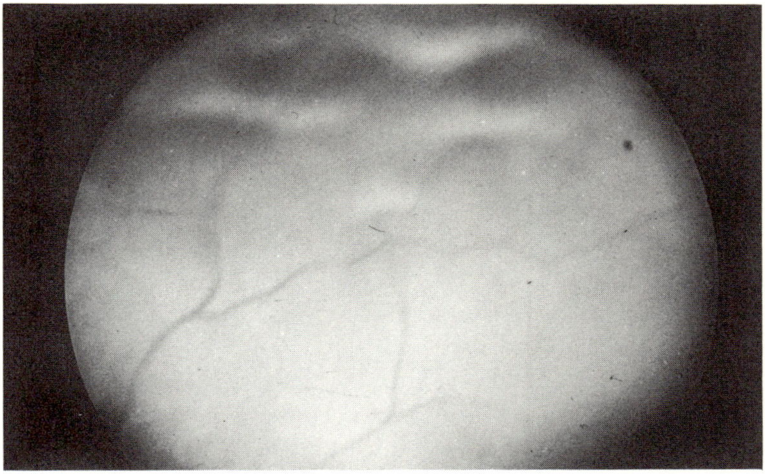

Fig **4.64**

lenses may improve close vision if the condition involves both eyes.

Tumors

Tumors of the choroid may be benign or malignant. Benign pigmented tumors called nevi are not serious. Malignant tumors, such as a melanoma, are serious. They may produce loss of sight and visual field defects, and they may spread throughout the body. Removal of the eye may be recommended even if 20/20 vision is present. If the malignancy has not spread, this may be a life-saving operation.

RETINA

Many systemic diseases leave their imprint by involving the retinal and choroidal layers of the eyeball. Local conditions in the retina or choroid may also impair sight.

Detached Retina

The retina becomes detached or separated from the vascular choroidal layer at the back of the eye just as wallpaper separates from a wall. If water is poured through a hole in the paper near the ceiling, the lower portion pulls away from the wall. Compare the appearance of a normal retina (1.4) with that of a detached retina (4.64).

WHAT CAUSES A DETACHED RETINA?

Overwork, worry, nervousness, eyestrain, or wrong glasses will *not* cause a detached retina. A vitreal-retinal adhesion at a weak spot in the retina usually pulls and makes a hole or break in the retina. The retina separates from the choroid. The detachment may be flat or elevated. Ballooned away from the choroid like a rolled up film separated from the camera, the elevated retina produces a poor picture. Some conditions predispose to a detached retina.

Fig **4.65**

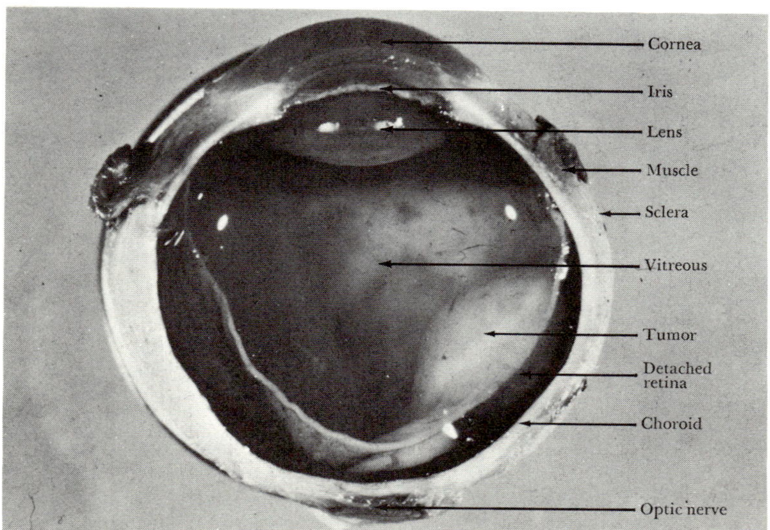

WHO IS SUSCEPTIBLE TO A DETACHED RETINA?

Persons with a high degree of myopia experience 30 to 40% of all retinal detachments; about 25 to 30% of these detachments are bilateral.

Degeneration, or breakdown of tissue due to aging processes, makes the retinal area thin and weak. However, the same changes of aging predispose to cataract formation. About 20% of all retinal detachments develop within two months to four years after cataract extractions, as the retina loses some of its support during this operation.

Injury from a fall that causes a blow to the eye or bump on the head may account for about 10% of all retinal detachments.

A tumor in the eye or previous inflammation in the retina and choroid also increases the risk of retinal detachment (4.65).

Hemorrhages, infections, or changes in the vitreous humor may cause it to shrink and pull on the retina. This tugging action may tear or break the retina in a weak spot. The fluid within the eye seeps through the holes and under the retina, ballooning it forward and separating it from the choroid, as water lifts and separates wallpaper from a wall after it seeps through a hole in the paper.

WHAT ARE THE SYMPTOMS OF A DETACHED RETINA?

If the film in a camera curls away from the back of the camera or is exposed to light, the picture will be ruined. The same thing happens when the retina is separated from the choroid. To function normally the retina must always be attached firmly to the choroid; if it is separated, poor vision or blindness will result.

A patient with a detached retina may experience blurred or distorted vision, lightning-like flashes of light, and sudden showers of cobweb-like black specks or strands. These vitreous floaters may be small or large, single or multiple; they may appear like string, thread, spiders, or coal dust. Other symptoms include the appearance of soot-like spots, a veil or curtain coming over the vision as the retina hangs down, or sudden loss of vision in the area of detachment. A person noting any of these symptoms should consult his ophthalmologist *immediately*.

The loss of sight is projected into the field of vision opposite from that in which detachment occurs: if the retina detaches from above, the patient cannot see objects below, and vice versa. The patient with a detached retina often sees better in the morning after bed rest; the fluid seeps back through the holes or tears in the retina, permitting it to flatten out. During the day the condition may worsen as fluid seeps back under the retina causing it to separate further.

HOW IS A DETACHED RETINA REPAIRED?

Detached retinas do not heal by themselves. Operative measures are required. If retinal detachment is diagnosed early, the patient's visual recovery can be good with new surgical techniques. Since most detachments occur in elderly people, the physician must weigh the overall picture in deciding whether to subject to an operative procedure an already upset, apprehensive, and depressed patient who has suffered a sudden loss of vision in one eye. Although the results of this procedure are not as dramatically successful as those of cataract extraction, some vision is better than none. If the decision is made to operate, the sooner the operation is carried out, the better. The longer the retina is detached from the choroid, the less likely the chances of restoring good useful vision to that eye.

Preparing the patient for operation. Using pictures and illustrations, the ophthalmologist explains to the patient just what a detached retina is, how it occurs, and what the surgeon will try to do to restore sight. The main objectives of the operation are to seal off the retinal hole or tear, to drain the subretinal fluid, and to reunite the retina to its normal flat position next to the choroid. The success of the operation depends on the nature of the detachment; the location, extent, size, number, and position of holes or tears in the retina; and other factors (vitreous changes, retinal degeneration, uveitis, injury, myopia, hemorrhage, and previous cataract or retinal surgery). Dilating drops are instilled into the eye to enable the doctor to see the back of the eye more clearly. He may spend several hours looking into the patient's eye and making maps or diagrams to help localize the holes or tears in the retina.

If the procedure is to be performed in the hospital, the patient washes his hair before going to the hospital and brings his toilet articles, bathrobe, slippers, and a radio. He will remain there from 2 to 14 days depending on the surgical technique used. If general anesthesia is to be used, relatives or friends usually arrange to sit with the patient for the first day or two after the operation.

The operation may be performed under local or general anesthesia. If general anesthesia is used, the patient feels only a small prick of the needle as it enters the vein of his arm, and then he goes to sleep. If a local anesthetic is used, he feels a fine stinging sensation at the side of his face and eye. Regardless of the type of anesthetic used, the operation is painless and lasts from one to four hours depending on the technique employed. Surgery is designed to "spot weld" the retina back into place. Various techniques may be used to produce scar formation, or adhesions, between the retina and the choroid.

Coagulation. With this technique adhesions are created by heating or freezing the tissue. Heating may be performed by electric diathermy. A tiny electric current applied to the outside of the eye surrounding the hole causes inflammation and coagulation of the tissue. As the healing process takes place with scar tissue formation, the retina is "glued" to the inner wall of the choroid. Heating the tissue to coagulate it is like vulcanizing a hole in an automobile tire.

Coagulation can also be carried out by a new and popular technique called cryopexy. A low-temperature cryoprobe produces adhesions of the retina to the choroid by freezing, rather than heating, the tissue. When detachments are originally flat (little or no subretinal fluid elevates the retina from the choroid) or when retinas have become flat after bed rest has been prescribed, a photocoagulator can be employed to seal off the retinal hole. This new instrument, which is used like an ophthalmoscope, flashes a tiny pinpoint of white or colored light into the patient's eye. The light coagulates and seals off the tear without subjecting the patient to surgery, external diathermy, or cryopexy. Another new device, the powerful argon laser, emits a pinpoint beam of high-energy red light. Some of these newer techniques, which are used when the detachment is flat or only slightly elevated, are performed in the doctor's office or on an outpatient basis.

Shortening the eyeball. The retina may not flatten after bed rest. Bands of tissue in the vitreous may be pulling the retina and lifting it away from the choroid. To insure a secure closure of the retinal hole and to enable the retina to come into contact with the choroid, the sclera over the hole may be modified surgically. The eyeball may be shortened by taking out a portion of the sclera, or by folding the sclera inward in a "buckle" operation. The sclera itself may be folded inward, or a small piece of permanently implanted plastic material may be attached to the outside wall of the eyeball to create an internal hump of sclera and choroid. Since the retinal hole lies on this "buckle," the retina has a better chance to be scarred (glued) to the choroid. During surgery, the eyeball is not taken out of the socket; it is gently rotated in the orbit. With this shortening technique the patient no longer has to dread lying flat in bed with sandbags to hold his head still for two weeks. He may sit up in bed the day after surgery, and he may be discharged from the hospital in five to seven days.

The shortening operation and a coagulation procedure may be performed at the same time. Although most retinal detachments respond to one operation, additional surgery is sometimes necessary. Since the retina is part of the central nervous system tissue, it cannot be transplanted from one eye to another, as can the cornea. If the retina is destroyed, no functional sight returns, regardless of how successfully the retina has been reattached.

Vitreous injection. The retina also can be made to approximate the choroid by injecting new vitreous fluid. This fluid, which may be saline or that obtained from a deceased donor's eye, is injected into the vitreous cavity. This creates extra internal pressure that pushes the retina closer to the choroid. Vitreous injection may be used in combination with other methods of retinal reattachment.

Postoperative care. After surgery, the eye usually is patched for three weeks until the stitches fall out, absorb, or are removed. After coagulation, pinhole glasses may be worn for several months. After the eyeball shortening operation, this may not be necessary, but the patient may wear sunglasses to avoid glare.

Favorable postoperative signs are wider sight (the patient sees with the whole eye rather than a part of it) and the sensation that the glob or blob that had obstructed his vision has disappeared. The patient may also experience an increased sensitivity to light. His vision may be hazy at first, but it improves with time and can even return to normal. After the dressings have been removed, drops or ointment are prescribed for the healing eye and warm compresses are applied several times a day with the eye closed.

After his operation the patient returns gradually to his daily habits and pastimes. Generally, he can resume the following activities after these periods of time:

- Shaving with a safety razor (an electric razor produces too much vibration)—four to five days.
- Taking showers or baths (with care not to wet the dressing)—three weeks.
- Reading—one month.

- Stooping, bending, and lifting (careful squatting is permissible earlier)—one month.
- Taking long automobile rides—one month.
- Returning to work—four to six weeks.
- Examination for glasses—three to six months.
- Washing the hair—two months.
- Engaging in athletic activities or contact sports—three months.

HOW SUCCESSFUL IS RETINAL REATTACHMENT?

The results of operations for retinal detachment are always more uncertain that those of a cataract operation. Retinal reattachment involves working with abnormal tissue that may have undergone degenerative changes. If the retina is reattached and holding firmly for three months, it is likely that it will be permanently attached. This is known as an anatomic success. If the visual function returns to normal, then functional success has been achieved as well.

Sudden Loss of Vision Due to Other Retinal Conditions

Conditions other than detached retina that may cause sudden painless loss of sight or blindness are hemorrhages into the vitreous accompanying arteriosclerosis, diabetes or hypertension, and occlusion of the central retinal artery or vein. A person developing sudden loss of vision should consult his physician immediately since diagnosis can be made only by the use of an ophthalmoscope.

Retrolental Fibroplasia

A much publicized retinal disease occurring in premature newborn infants, this condition has been found to result from an overadministration of oxygen to these "blue babies" in an attempt to save their lives. Retrolental fibroplasia may lead to blindness, and there is no known therapy to reverse already established changes. Since the cause of the disease is now known, measures have been taken by physicians caring for newborn infants to make retrolental fibroplasia a preventable disease.

Degenerative or Aging Changes

CENTRAL LOSS OF VISION

Poor circulation or aging processes in the elderly may cause changes in the central part of the vision, the macula lutea. Macular degeneration impairs central or direct sight, but not peripheral or side vision. Although a person cannot see an object when he looks directly at it, as in reading, he can see it by looking out of the corner of his eye. A patient with central macular degeneration will not go totally blind, but he will have difficulty seeing objects clearly and acutely. Treatment to reverse this condition is still forthcoming.

PERIPHERAL LOSS OF VISION

Premature aging of the retina with pigmentary changes at its outer portion may occur with syphilis or retinitis pigmentosa. Retinitis pigmentosa, a recessively inherited pigmentary degenerative disease, progresses slowly with age. Starting at the periphery of the retina, it may reduce night vision and narrow side vision. These patients, therefore, should not drive a car or be on the streets alone at night. Side vision may narrow to tubular or gun barrel vision as in glaucoma patients, but central sight may remain intact for some time. There is no known treatment to reverse the course of this condition. Telescopic or magnifying lenses will sometimes improve central and peripheral sight in degenerative retinal conditions.

Tumors

Angioma and retinoblastoma are typical tumors of the retina. Angioma, a rare blood vessel tumor that may produce loss of sight, may be destroyed by coagulation. Retinoblastoma, an extremely malignant tumor, usually occurs in children. All children with a crossed eye should be examined to be certain that this tumor is not the cause of the strabismus. The child's eye may look like a cat's eye because of a white or yellowish reflex from the pupillary space. This cat's eye reflex is the result of the tumor's projection as a polypoid mass into the vitreous. If one eye is involved, enucleation is usually recommended; these tumors often spread to the rest of the body if not discovered and treated early. If both eyes are involved, radiation or injection of chemicals into the bloodstream may be tried if any sight is present in the second eye; otherwise both eyes may have to be removed in an attempt to save the child's life.

VITREOUS

The vitreous, or the jelly-like transparent mass behind the lens, fills four fifths of the eyeball. It has no blood supply, but it may be involved with disease by the structures that surround it. Bleeding from the retina into the vitreous may cause a vitreous hemorrhage with sudden impairment of sight. Infections from the retina or uveal tract may produce a hazy vitreous, with loss of sight. Sudden detachment of the vitreous can create spots, cobwebs, and specks of pigment floating in the vitreous like the stars seen when the eyes are pressed or rubbed. Spiders and flashes of light or rings in the visual field are similar to those seen with a detached retina. Central sight usually is good, but the patient should be examined by his ophthalmologist to rule out the possibility of a detached retina.

Floaters

Single floaters in the form of threads, hairs, or spots do not always indicate retinal or vitreous detachment. The sometimes annoying visual phenomenon of vitreous floaters is best explained by analogy. Visualize a cut on your hand. As the cut heals, a scab forms. If this scab were put into a glass of water and held before your eyes while the glass was being rotated back and forth, the scab would float past your line of vision and you would see it.

Vitreous floaters may be flecks of pigment that have dislodged from the retina in nearsighted eyes or in patients with healed chorioretinitis, or they may be blood clots or exudates (red or white cells and fibrin). They may also represent vitreous degeneration resulting from changes in the consistency of the vitreous as seen by reflected light. Floaters are visible because they interfere with the light passing into the eye. All patients who have vitreous floaters do not necessarily have incipient retinal detachments, especially if the floaters are few in number, but a retinal detachment always should be ruled out by an ophthalmologist. Proper medical treatment may reduce the size of bothersome floaters.

Hemorrhage

Although large hemorrhages in the vitreous may absorb, they may leave bands of scar tissue that extend from the vitreous to the retina. This condition, which is called retinitis proliferans, may cause loss of vision or predispose to a detached retina. Vitreous normally has a gelatinous consistency in young persons, but it becomes more fluid in older people. This transformation in the vitreous also may occur with high myopia, degeneration, or with inflammatory diseases involving the retina or choroid.

OPTIC NERVE

If this tube-like extension from the brain to the eye is injured, the nerve fibers never completely regain normal function. If it is completely severed or destroyed, sight is totally or permanently lost in that eye.

Various conditions may involve the optic nerve. Pressure on the optic nerve from glaucoma can impair its blood supply and nutrition. The nerve loses function and the eye loses sight. Obstruction of the central retinal artery or vein, both of which run through the optic nerve, can produce sudden loss of sight. Hemorrhages or tumors may also involve the optic nerve and impair vision. Inflammations involving the optic nerve (optic neuritis or papillitis) either before or after the nerve enters the eyeball can cause loss of central sight. Retrobulbar neuritis, which is an inflammation of the optic nerve behind the eyeball, may accompany multiple sclerosis or toxic conditions and may produce sudden impairment of central vision. Toxic conditions caused by ingestion of methyl alcohol or by excessive intake of tobacco can injure the normal function of the optic nerve. These conditions produce a central blindness called alcohol or tobacco amblyopia.

Papilledema is swelling of the optic nerve. Although it produces no gross symptoms, this swelling may indicate the presence of hypertension, inflammation, or a brain tumor or other serious lesion of the brain. Since papilledema can be detected only with an ophthalmoscope, a complete eye examination is imperative.

Optic atrophy is loss of function or death of the optic nerve due to the effect of pressure or inflammation. Characterized by a reduction or loss of vision that is usually permanent, this condition is not reversible and does not respond to any form of treatment.

VISUAL FIELDS

The visual field embraces total sight. Loss of peripheral

or side vision may be caused by diseases of the retina, choroid, optic nerve, or the brain itself. Lesions within the eye (detached retina, hemorrhages, tumors, or chorioretinitis) produce visual field loss. Diseases within the brain (hemorrhages, tumors, or inflammations) often cause peripheral visual field defects. In fact a visual field loss may be the first sign of many of these diseases. Since visual pathways run from the eye and travel through the brain, different lesions in the eye or brain may produce various field defects that may be detected by perimetric studies.

In glaucoma the peripheral field of vision may be constricted first without the patient being aware of it. Blind areas around the normal blind spot, as well as other changes, may be detected by visual field studies.

A brain tumor may impair the temporal parts of the visual fields in each eye. This loss of side vision, or bitemporal hemianopsia, is similar to that experienced by horses wearing "blinders." A tumor may also produce blindness on one half of the field of vision; the patient may not see anything on his right side with either eye.

A detached retina causes a visual field loss corresponding to the area of retinal detachment.

The excess use of tobacco or alcohol may cause loss of central but not peripheral vision.

Injury or a tumor that presses on the optic radiations anywhere in the brain—from the chiasm to the occipital lobe—may produce a loss of half of the visual field in each eye, called hemianopsia. The patient can only see objects on the left or right side of the midline.

The course and progress of these diseases can be followed by visual field studies. (See pages 48–51.)

LENS

Cataracts

A cataract is an opacity or cloudiness of the normally

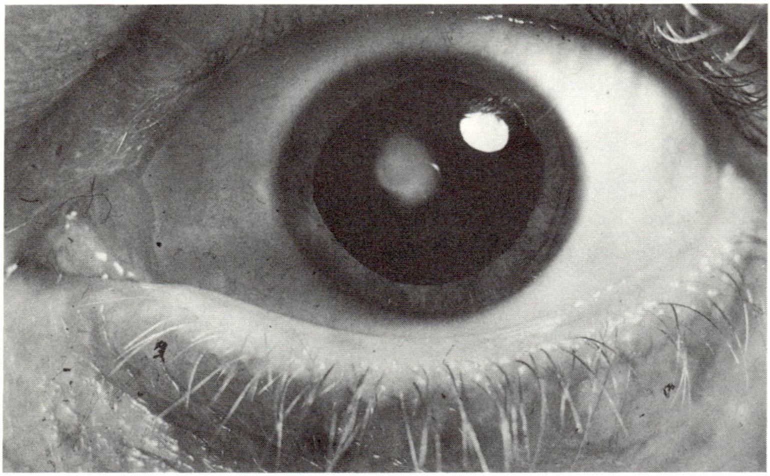

Fig 4.66

transparent, crystalline lens. It may appear in the form of small dot-like opacities or as one large opacity involving the entire lens. Normally the lens has a slightly yellowish tinge. With cataract formation it often takes on a deep-yellow or amber tint, which subdues the colors blue and yellow as they come through the lens. However, the lens may become brown, gray, or white with cataracts of different types. By cutting off clear rays of light entering the eye, the cataract impairs vision.

A cataract may be compared to a window. If the window is clean, everything on the other side of the glass will appear sharp and clear. If you smear soap on the window, everything on the other side will appear hazy. But soap, unlike paint, can be wiped off. If you smear paint over the glass, everything will be fuzzy or cloudy or not visible at all. This is the way objects appear to a person with a cataract. To see better, the painted glass window must be replaced with clear, clean glass. The cataract is like paint smeared on the lens of the eye and can be removed surgically.

The cause of cataracts is still a matter of conjecture, but the mechanism of their development is fairly well understood.

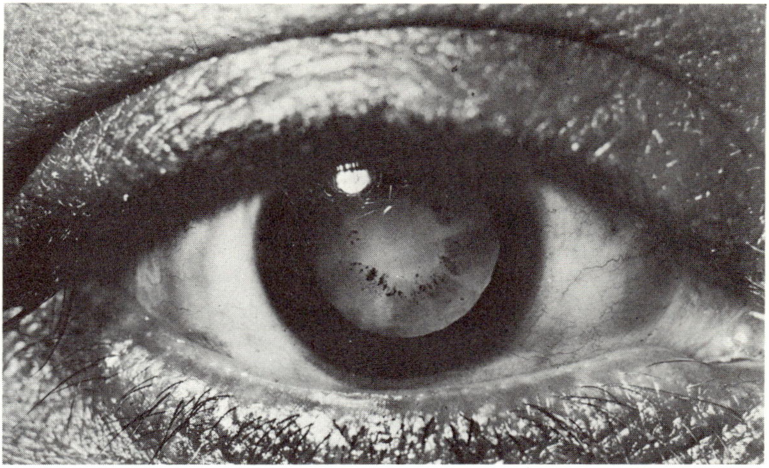

Fig **4.67**

Some cataracts are present at birth; others are the result of various local or systemic conditions. There are several common types of cataracts.

CONGENITAL CATARACT

This type of cataract, which is due to faulty development of the fetus, is more prone to develop if the mother has had measles during the first three months of pregnancy. To detect this condition every child should be examined carefully during the first six months of life by his pediatrician or family doctor. If a cataract is discovered, the child should be referred to an ophthalmologist. If the congenital cataract involves only part of the lens and does not affect vision (4.66), no surgery is required. These cataracts are usually stationary and do not progress. If the cataract involves the entire lens (4.67), surgery is required to restore sight.

Since central visual acuity (macular vision) develops best during the first years of life, surgery should not be delayed if vision is markedly impaired. The earlier the cataract is removed, the greater chance the child has to develop good vision with cataract glasses. If congenital defects other than the cataract are present, however, normal sight will not be regained fully despite the success of surgical treatment.

SENILE CATARACT

This is the most common type of cataract. It usually occurs in people past 50 years of age due to a physical aging process, poor lens nutrition, degeneration, or an inherited tendency for cataract formation. There are three main types of senile cataract.

Nuclear. This hardening of the center of the lens (4.68a) may produce pseudomyopia and "second sight." A person who required a farsighted glass to see or read now needs no glasses at all or a nearsighted corrective lens; therefore, he thinks his sight is getting better. The cataract is usually brown in color, in contrast to the normal yellowish or gray lens, and it is slow in developing.

Cortical. This cataract appears in the outer layer of the lens (4.48). It may be slow growing, or it may grow rapidly

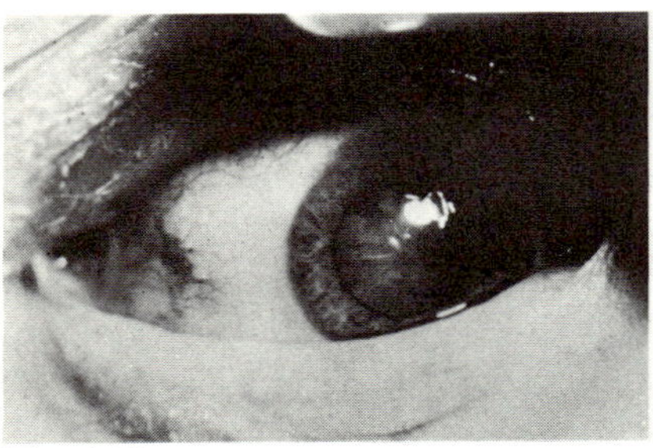

Fig 4.68a

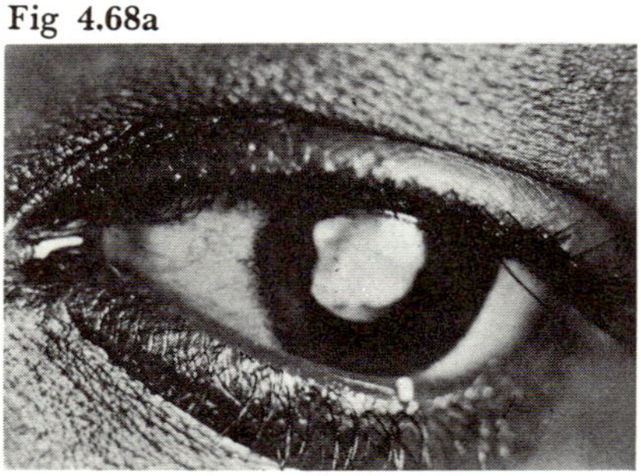

Fig 4.68b

and cause loss of sight. It is usually white. Both its development and its appearance are due to the absorption of aqueous humor by the lens. Since the cortical cataract may swell, pushing the iris forward, it may cause secondary glaucoma. Therefore, this cataract should be removed when it is mature not only to improve sight, but to prevent glaucoma.

Posterior (subcapsular). Since this cataract appears on the back surface of the lens, it may not be visible to the naked eye, but be present instead as a haze in the black pupillary space. The posterior cataract may progress slowly or rapidly, but it will often impair vision early because of its central location.

Since any of these three types of senile cataracts diminish vision only slightly as they begin to form, they are called incipient cataracts. As cloudiness progresses and vision becomes more impaired, the cataracts are called immature cataracts. When degeneration of the lens is complete and vision is very much impaired, the cataract is ripe or mature.

TRAUMATIC CATARACT

Injury to the eye from a blunt or penetrating instrument or object causes rupture of the lens capsule (4.68b). When this occurs, the lens imbibes, or takes in, aqueous humor, becomes cloudy or opaque, and interferes with sight. The cataract must be removed to restore sight. Final visual results, however, depend on the state of the other ocular structures. If a detached retina is present after the traumatic blow, the detachment might only be detected after the cataract is removed. To restore sight the retina would have to be reattached. Dislocation of the lens may be the result of injury. If this dislocation causes secondary glaucoma or interferes with vision, the lens may have to be removed.

COMPLICATED CATARACT

This form of cataract may be associated with such systemic diseases as diabetes, tetanus, calcium deficiency, hyperthyroidism, atopic dermatitis, or myotonia atrophica. Chemical agents, such as ergot or dinitrophenol, also may cause a complicated cataract. Radiation cataracts can result from atomic emanations. X-rays, plus infrared rays from intense heat, may destroy lens cells, producing the heat radiation cataracts seen in glass blowers or steel puddlers. Radiation cataracts have now almost been eliminated due to safety precautions.

Cataracts also may develop as a complication of such local eye diseases as uveitis, iritis, chorioretinitis, glaucoma, tumors, detached retina, injury to the eye, a high degree of myopia, and retinitis pigmentosa. Since this form of cataract may occur in patients 20 years of age or younger, it should be removed when it impairs the vision enough to interfere with the person's occupation.

WHAT ARE THE SYMPTOMS OF CATARACTS?

Depending on the type of cataract, one or more of the following symptoms might be observed: blurry, smoky, fuzzy, cloudy, or foggy vision; dazzling stars or rays emanating from lights; double vision in the eye with the cataract; colored rings or halos around a light; or black spots or shadows that appear to move when the eye moves if the cataract is in the center of the lens. The patient may also find that reading matter must be held more closely or that he can never get the light for reading quite bright enough. He may need frequent changes of glasses. His distant vision is clearer than his close vision because the pupil is more dilated as the visual line of sight passes around the lens opacity. He may see better in the evening when the pupil is dilated than in daylight or artificial light when the pupil is constricted. A patient exhibiting these symptoms does not necessarily have cataracts, but he should have his eyes examined.

MISCONCEPTIONS ABOUT CATARACTS

If you have a cataract, you will go blind. A cataract is one eye disease that can be remedied by a painless operation to restore sight. Of course, the patient's activities will be limited by his poor vision until the cataract is removed from at least one eye. He should not work if his occupation is hazardous. He should not drive an automobile at night. He should not cross streets alone in the dark or in the dazzling sunlight.

Cataracts are a skim or skin growing over the eye. This is not true. The cloudy lens lies inside the eye behind the black pupillary space.

You will not get a cataract if you do not use your eyes too much. This is also a misconception. Prolonged use of the eyes, with or without cataracts, will not harm them. However, correct glasses, good light, and proper reading habits should be used at all times. If fatigue develops, the eyes should be rested momentarily.

Cataracts spread from one eye to the other. Frequently a cataract progresses more rapidly in one eye than in the other, or only one eye is involved by a cataract. However, a cataract cannot spread from one eye to the other since it is not the result of an infectious or malignant process.

The presence of cataracts means death to the lens tissue. A cataract is similar to gray hair. Just because the hair is gray does not mean that it will fall out or die. A cataract, like graying of the hair, is an aging process.

If your parents had a cataract, you will have one. Since congenital cataracts may be inherited in a dominant, recessive, or sex-linked recessive fashion, there may or may not be a tendency for transmission from one generation to the next. Presenile cataracts (those occurring in early or middle life) may be present in several members of a family.

A cataract should be ripe, or the lens completely opaque, before it can be removed. In the past this was true, but new surgical techniques and instruments have changed the approach to cataracts. In the ripe or hypermature state, cataract surgery is more difficult, and complications arising from toxic products in the lens may impair the health of the eye. Why

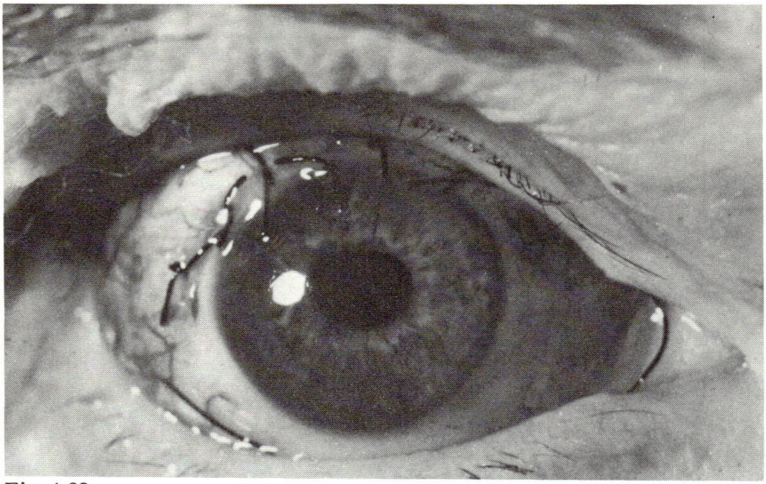

Fig 4.69

wait until a person is blind or incapacitated if he can be restored almost immediately to a normal self-sufficient life? Surgeons now operate to remove a cataract when 50% of the vision has been impaired and the patient is unable to see to perform his work efficiently. The operation is also performed when the better eye is unable to serve the patient satisfactorily, or when the patient is no longer happy with his present state of sight.

The following case report illustrates another reason why the ophthalmologist no longer delays removal until that cataract has become ripe:

> A 39-year-old bus driver was examined because he complained of hazy vision. His visual acuity was: right eye 20/50, left eye 20/50. Examination revealed early posterior complicated cataracts. Because of their central location, the cataracts prevented him from seeing well enough to drive. If he waited until the cataracts became ripe and his vision deteriorated, he would be without a job and unable to support his wife and three children.

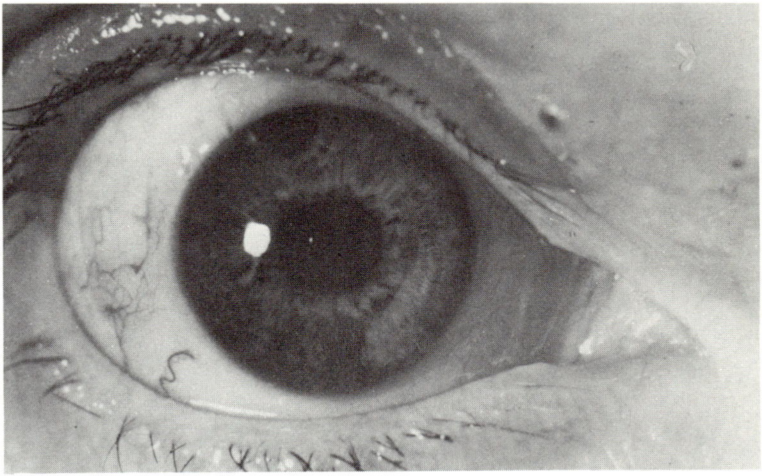

Fig 4.70

He was advised to have a cataract extraction in one eye. The operation was successful (4.69). One month later with a cataract lens he was reading 20/25. Surgery was performed on his other eye one month later. Vision in this eye was similarly improved, and the man returned to his work driving a bus four months after his first operation. One year later with permanent cataract and/or contact lenses (4.70) his vision was 20/15 in both eyes.

Rapidly improving vision requiring frequent changes of glasses or no glasses at all is a good sign. With the development of nuclear cataracts, the patient notices that frequent changes of glasses are required to improve his vision. He may see better for close work by using a weaker lens or even without his glasses. This "second sight" is in a sense false security. As the cataract progresses, vision becomes gradually poorer and no lens will give improvement. The only therapy for the patient who has reached this stage is cataract extraction.

Age is a deterring factor in cataract extraction. Patients from 6 months to 105 years old have been operated on for cataracts. The question is not age, but how much the patient wants to see again. This point is illustrated in the case report of this 105-year-old patient:

> This gentleman wanted to be operated upon so that he would "be able to read the *Wall Street Journal.*" His cataract extraction was successful: with cataract lenses he had 20/20 vision and Jaeger i at 14 inches, enabling him to read the daily quotations as well as "look at the pretty girls."

Poor general health makes cataract extraction impossible. At one time poor health was a contraindication to surgery. Since the operation is painless, without shock to the system, and involves relatively little blood loss, this is no longer true unless the patient is gravely ill. This unusual case report describes the restoration of health, as well as sight, by cataract removal:

> A 72-year-old Chinese man weighed 68 pounds when he was first examined. He had bed sores and was being fed intravenously. He had been lodged in a side room on the medical ward of a large hospital until one of the medical residents noticed that he had cataracts and had him transferred to the eye service.
>
> Cataracts were removed successfully from one eye and then, two weeks later, from his other eye. He was given temporary cataract glasses the next week. After he could see again, he started to sit up in bed. He then started to eat by himself. When the doctors would make ward rounds, he clasped his hands and bowed up and down in an attitude of thankfulness. His weight increased to 110 pounds in the following weeks. Not only did he eat his own food, but he got out of bed and stole food from other patients and from the food cart.
>
> Many old Chinese feel that when sight is gone, life is ended. Since this man could see again, he had a new life.

HOW ARE CATARACTS REMOVED?

Preparing the patient for surgery. Before going to the

hospital, the patient is told that the visual outcome depends upon three things: a successful operation; a good patient who does not pull off his dressing or stick his finger in his eye, vomit, sneeze, or fall out of bed; and a functioning retina and optic nerve. The state of the retina and optic nerve cannot always be determined before surgery because these structures are hidden behind the cloudy or opaque lens. If optic atrophy or retinal degeneration is present, the final visual result may not be as satisfactory as both patient and doctor desire.

The patient washes his hair before going to the hospital, as it probably will be six or eight weeks until this can be done after the operation. Antibiotic eye drops to be instilled into the eyes several days before surgery are usually prescribed to help prevent infection. If an infection is present, surgery is postponed. The patient is advised to be careful to avoid catching a cold. Since a head cold and cough may cause postoperative complications, it is a contraindication to surgery. The patient should bring toilet articles and gowns, which do not have to be pulled over his head, with him to the hospital. Nurses, friends, or family members should plan to sit with the patient 24 to 48 hours after surgery to see that he does not injure his eye in his partially anesthetized state. The operation is described briefly to the patient; he is shown the location of the lens and an actual lens that has been removed.

Hospital routine. If surgery is performed in the morning, the following routine is observed: The patient washes his face the night and morning before surgery. A sleeping pill is given the night before surgery. The operation is quite painless, and there is no shock to the system. The patient is put in twilight sleep, and the eye is anesthetized locally to keep the eyelids and muscles to the eye motionless. General anesthesia is used for apprehensive patients and for children.

Operative procedure. During the operation, the eye is not lifted out of the socket onto the cheek. The eyelids are open and the eye is in place. An opening is made in the cornea and sutures are inserted that will close this wound after the cataract has been removed. A hole is then made in the iris. This hole may be peripheral and very small, preserving the round pupil, or it may be a complete key-hole shaped wedge. By allowing a free flow of aqueous humor from the posterior to the anterior chamber this operation prevents the iris from being pushed into the wound, causing secondary glaucoma. The lens is then removed by capsule forceps, an air suction technique, or a cryosurgical unit that freezes to the lens.

If the whole lens and the contents within the lens capsule are removed, the operation is intracapsular cataract extraction. This technique usually is performed on adults.

If part of the contents of the lens is removed and part of the lens material and capsule is left in the eye, the operation is extracapsular cataract extraction. This technique has been used in the past in performing cataract surgery on children. Intracapsular lens removal has been facilitated by the use of the enzyme alpha-chymotrypsin, which weakens the zonules or ligaments that support the lens and aids in its removal.

Fig 4.71

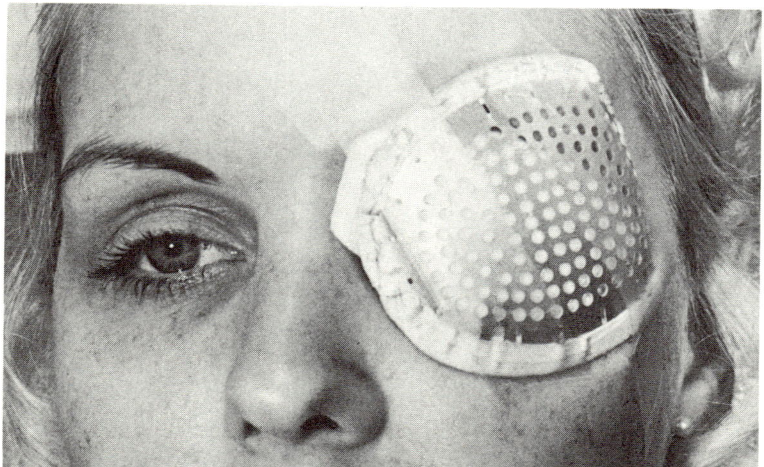

After all or part of the lens is removed, the sutures are tied, sterile fluid and air may be injected into the anterior chamber, and the wound is closed securely with an air-tight closure. The eye is then treated with antibiotic ointments or solutions. The operation generally lasts 30 to 60 minutes. Usually only the eye that has been operated on is bandaged; it is then covered with a protective metal shield (4.71).

CATARACT REMOVAL IN THE OPPOSITE EYE. Two weeks to several years may pass before the cataract is removed from the other eye. Only rarely does the ophthalmic surgeon remove both cataracts at one time. A few patients may be satisfied with the successful result on one eye and not have the other cataract removed; others are so pleased that they beg to have the operation performed on the other eye. It is reassuring to the patient to have two seeing eyes, or one in reserve.

CATARACT REMOVAL IN CHILDREN. The operation for the removal of congenital cataracts in children is different from that in an adult. It is difficult to remove the lens completely within its capsule because the lens adheres to the surface of the youthful gelatinous vitreous. Therefore, the lens capsule is opened surgically, permitting aqueous humor to come in contact with the lens structure. In the past the cataract material was left to absorb so that the central pupillary space would clear. If absorption was not complete, additional surgery was required to produce a clear central opening. Today a newer technique permits aspiration of the entire lens content. Most recently a procedure called phakoemulcification is being used in certain eye centers. The lens is disintegrated and aspirated out of the eye with an instrument that could revolutionize cataract surgery.

DIAGNOSTIC AND COSMETIC CATARACT REMOVAL. Rarely cataracts are removed not only to restore sight, but for diagnostic as well as cosmetic purposes. If an eye is partially blind because of a previously diagnosed retinal disease and has an unsightly white pupillary area, the cataract may be removed to study the fundus as well as to improve the appearance of the eye. However, probably no visual improvement will follow. An artificial plastic contact shell, which bears an exact likeness to the other eye, may be used for cosmetic purposes.

Care after surgery. Depending on his surgeon and on his postoperative course, the patient is permitted to sit up in bed the first day, get out of bed on the second or third day, and leave the hospital after 5 to 14 days. Of all surgical procedures cataract extraction is one of the easiest and most pleasant for the patient. Thanks to new surgical techniques, sandbags are not used and the patient no longer has to lie rigid in bed for two weeks. Because of sharper needles and finer sutures, the wound can be tightly closed with many sutures. Therefore, the patient has more freedom of movement

AFTER YOUR CATARACT OPERATION

Your wound is healed, but it will not be firm enough to stand much pressure for three months.

1. Continue to be as careful at home as you have been in the hospital the past few days and have very few visitors.

2. Avoid closing the eyes tightly. One often closes the eyes tightly when laughing, talking, sneezing, coughing, or if they feel irritated. At these times you should be particularly careful not to close your eyes tightly.

3. Avoid stooping, straining, lifting, bending, or reaching.

4. Change eye dressing daily.

If there is any secretion, gently wipe off the lids with cotton moistened with tap water, but *avoid exerting pressure* on the eye, particularly the upper lid.

You may be given drops or salve to use in your eye. If so, follow these directions carefully. To instill drops in your eye, open it, *look up,* and have someone *gently* pull the lower lid down and instill a couple of drops. To put salve in the eyes, do the same as for the drops except that you express from the tube about one-half to one inch of the salve inside the lower lid. *Do not touch the upper lid.* Then apply an eye pad and tape.

Next, apply a metal shield which will be worn at night for another two months. Apply the shield with adhesive tape one-half inch wide.

5. You may do near work, such as reading or sewing, and you may watch television, play cards, wipe dishes, etc.

6. Do not have your hair washed for two months after the date of surgery. A hair trim may be performed after one month.

7. Men may shave with lather and a new blade—no electric shavers.

8. You may use a sponge bath for three weeks, then a tub or shower *with help.*

9. You may wear dark glasses with or without your own glasses if you are in bright light. A short walk or drive may be taken.

10. The sutures may give a feeling of something in your eye but do not close it tightly. They will be removed three to four weeks after surgery.

11. If you have sharp pain in your eye, or any unusual symptoms, call your doctor.

after surgery. If he were not so well sedated, he could actually get up and walk away from the operating table. He can read, play cards, watch television, and take short walks as early as the third to fifth day after surgery. However, caution is required in combing the hair (gentle brushing may be permitted), stooping, bending, and lifting for six to eight weeks after the operation. A complete list of instructions to cataract patients leaving the hospital appears on page 136.

If sutures that hold the eye in place after cataract removal are not buried, they may produce a scratchy sensation. This is relieved by aspirin and/or by keeping both eyes closed. The eye is dressed daily, or every other day, at which time antibiotic medication is instilled. The sutures, which remain in place for from 20 to 28 days, are then painlessly removed in the doctor's office. If catgut or buried silk sutures are used, there is no need for removal.

HOW SUCCESSFUL IS CATARACT REMOVAL?

With a better understanding of cataracts and modern surgical techniques cataracts should no longer be feared by anyone. If the patient's occupation involves only light desk work, he may return to work one month after the operation. Even though cataract extraction is a delicate surgical procedure, hospital records reveal that over 95% of all cataract operations are successful. A good visual result will be obtained if the patient is cooperative, before, during, and after surgery, and if the fundus or interior of the eye is normal. However, postoperative hemorrhage, infection, or glaucoma may interfere even with successful surgery.

After the dressings are removed, but before glasses are prescribed, the patient can often see better than before surgery, even though his vision may be poor since the focusing power of the eye has been removed. His peripheral vision as well as his entire field of vision is increased; he can now perceive large objects, even without a cataract lens. Since his eye no longer has a color filter, the sky is a deep Wedgwood blue. An eye with a missing crystalline lens is called an aphakic eye. The vision of this eye can be improved by a cataract glass that compensates for the absent lens.

WHAT ARE CATARACT GLASSES?

Temporary cataract glasses are prescribed anywhere from one week to six months or longer after surgery. Later permanent glasses are prescribed. These glasses must be worn for the rest of the patient's life. If he has had a successful cataract operation on one eye and has 20/20 vision in his other eye, he may have difficulty using both eyes together with a thick cataract lens on one eye and a thin lens on the other eye. The spectacle lens for the eye that has been operated upon produces a magnification ranging from 20 to 25%, in contrast to the lens for the other eye. The patient is usually forewarned of this difficulty before surgery. To prevent this state of rivalry between the eyes, glasses are prescribed to correct the vision in the better eye and blur the vision in the poorer eye. The patient is assured that his eyes will work together after both eyes have been operated upon. A contact lens, worn from four to six months after surgery on the aphakic eye, often will alleviate this difficulty and help prevent double vision.

Many improvements have been made in the thick forward cataract lenses. The use of aspherical lenses now permits sight through the entire lens, rather than through just the center. The adoption of plastic lenses makes the glasses feel lighter and appear less thick. Patients with new cataract lenses require much reassurance; learning to rejudge distances may be difficult. Steps or curbs look closer to the person, and he may have trouble setting a glass or book on the table or grasping a bannister. Straight lines may appear distorted and doorways may appear narrower. Since gauging size and distance is a function of the brain, it may take three to six months to adjust to this new experience.

Intraocular acrylic lens. Another modern advancement

is the insertion of an intraocular acrylic lens into the anterior chamber, between the iris and the cornea. With the insertion of this lens no spectacle or contact lens is required; the lens replaces the cloudy lens that has been removed. Although the use of this lens has not been fully accepted in this country, the intraocular acrylic lens may prove to be a remarkable advancement in the treatment of cataracts.

TREATING CATARACTS WITHOUT SURGERY

Vitamin C, in the form of ascorbic acid tablets or lemon juice or certain eye drops, dilates the blood vessels and improves the nutrition to the anterior segment of the eye. It may have some slight value in deterring the progress of a cataract. However, once a cataract is formed, no drugs, drops, or injections will cause the cataract to dissolve or disappear. The only cure for a cataract that interferes with vision is to remove the cataract through surgery.

GLAUCOMA

Glaucoma, or increased pressure of the fluid within the eyeball, is known commonly as "hardening of the eye or globe." Glaucoma is not an infection or a tumor. It should not be confused with trachoma, which is an inflammation, or glioma, which is a tumor. Many believe that a person with glaucoma will go blind. This is not altogether true, especially if the condition is diagnosed early before any vision has been lost and if effective treatment is instituted. While the ravages of glaucoma are not reversible, further damage to sight can be prevented. If, however, this condition is not diagnosed and controlled early, a person could lose his sight. Glaucoma might be compared to a rock lying on a bed of green grass. If the rock remains for a short period of time and is then removed, the grass will be a little crumpled, but will still remain green and viable. However, if the rock lies for a long time, the underlying grass will be destroyed. If the stone is removed before the grass is permanently destroyed, some pieces of the grass will still function and live. The retina and optic nerve fibers in the eye, which would be similar to the grass, will still survive if the pressure within the eye is relieved before the function of the structures is completely destroyed. The sight that is already lost cannot be restored, just as the dead grass will not come back to life.

Since glaucoma is often discovered during routine eye examinations when the pressure in the eye is checked with a tonometer, adults over 40 should routinely have eye examinations, which include the checking of pressure, every one to two years.

What Causes Glaucoma?

To understand how pressure in the eye increases, the eye may be compared to a bathtub (4.72). Water comes into the tub through the faucet and goes out through the drain. If the rate of inflow is the same as the rate of outflow, the water will not rise in the tub. However, if the water comes in faster than it leaves, water will rise in the tub. If debris clogs the drain, the water will not be able to leave the tub as fast as it is coming in and will remain to rise in the tub. If a rubber sheet is placed over the tub, the rising water will hit this sheet. As the pressure of the water builds up, the rubber sheet will start to bulge and may break. This may be prevented by turning off the faucet, by unplugging the drain, or by punching a hole in the side of the tub to let the water out.

Like the bathtub, the eyeball contains fluid. Aqueous humor is constantly flowing into the eye from the ciliary body (the faucet) into the anterior chamber (the tub), located in front of the lens and iris and behind the cornea. Aqueous humor exits at Schlemm's canal (the drain) located in the angle formed by the iris and the cornea. The fluid then drains

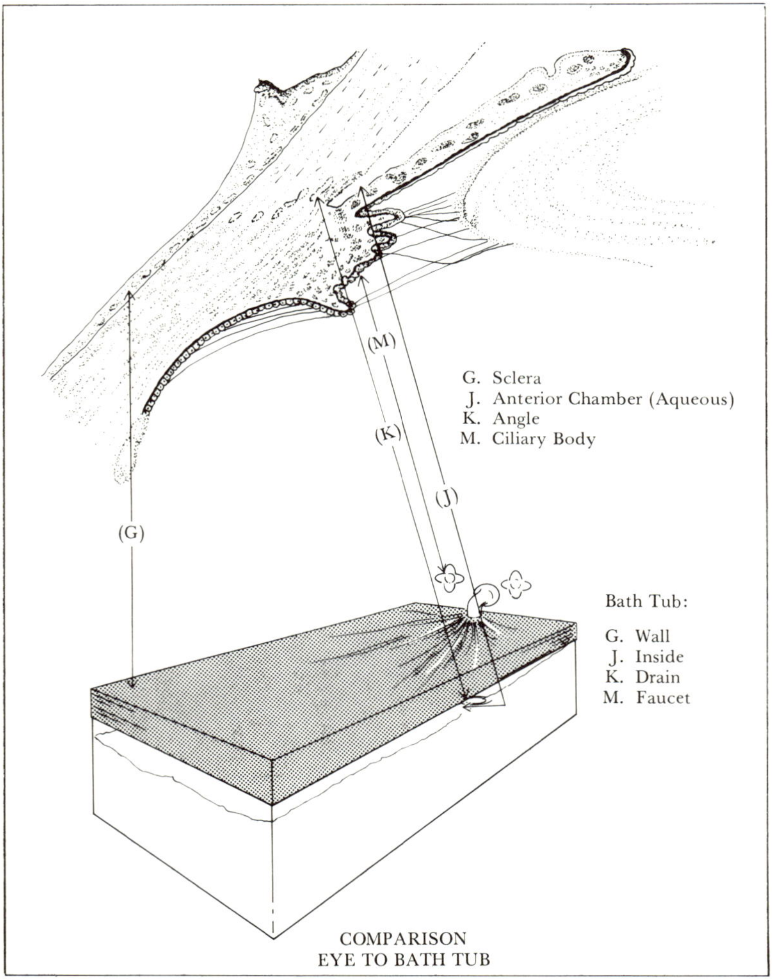

Fig 4.72

into the bloodstream via aqueous veins in the sclera. If aqueous humor enters the eye at a faster rate than it leaves Schlemm's canal (drain) or if it leaves the eye at a slower rate than it is being produced by the ciliary body (faucet) due to the presence of pigment in the angle (debris in the bathtub), pressure will build up in the eye and glaucoma will result. Glaucoma can be controlled by decreasing the amount of fluid coming into the eye, by increasing the outflow of fluid, or by performing a filtering operation to allow the aqueous humor to escape through some channel other than Schlemm's canal.

PRIMARY GLAUCOMA

Chronic, simple, or wide or open-angle glaucoma. Primary glaucoma is called "the thief of sight." Since it is a sneak thief, a person usually is unaware that he has glaucoma unless it is discovered during a routine eye examination when the intraocular tension is found to be elevated. Even then it may be too late; his vision may be markedly impaired. The patient may require frequent changes of his glasses or experience intermittent blurred or hazy vision. When the condition is finally diagnosed, an attempt is then made to control the pressure through medication or surgery to preserve the vision that is still present; however, this treatment cannot restore vision that has already been lost. Since this form of glaucoma occurs in people 40 and over, anyone in this age group should have a tension determination every one or two years as part of his routine eye examination.

Acute, congestive, narrow or partially closed-angle glaucoma. This type of glaucoma is characterized by intermittent pains in the eye that occur when the pupils are dilated—at night, in a movie or while watching television, while sleeping, or after excitement. The pain may be relieved by looking at a bright light or at the sun. Characteristically acute glaucoma can also produce a sudden onset of a severely painful red eye, loss of vision with a dilated pupil, aching of the face and head, hazy vision, and nausea and vomiting. This condition may be treated without too much visual loss. Under proper medication or surgery, it may remain controlled for the rest of the patient's life.

With either type of glaucoma a person may see halos or

rainbow-colored rings around lights, as if he had stepped out of a shower with water in his eyes. This phenomenon is due to the increased amount of fluid in the cornea. Other symptoms or signs are constriction or narrowing of side vision, small blind spots, blurred or foggy vision, intermittent eye aches, and frequent changes of glasses, which do not improve vision.

Since primary glaucoma usually is bilateral and may be more advanced in one eye, both eyes should be watched constantly and treated to control elevated tension. During a glaucoma examination the eyes are examined with a gonioscopic prism (4.73). This device, which is applied to the anesthetized cornea like a contact lens, enables the ophthalmologist to see the corneal-scleral-iris junction and to observe whether the angle is wide (deep) or narrow (shallow), or opened or closed.

SECONDARY GLAUCOMA

Secondary glaucoma may occur in a normal eye or with

Fig 4.73

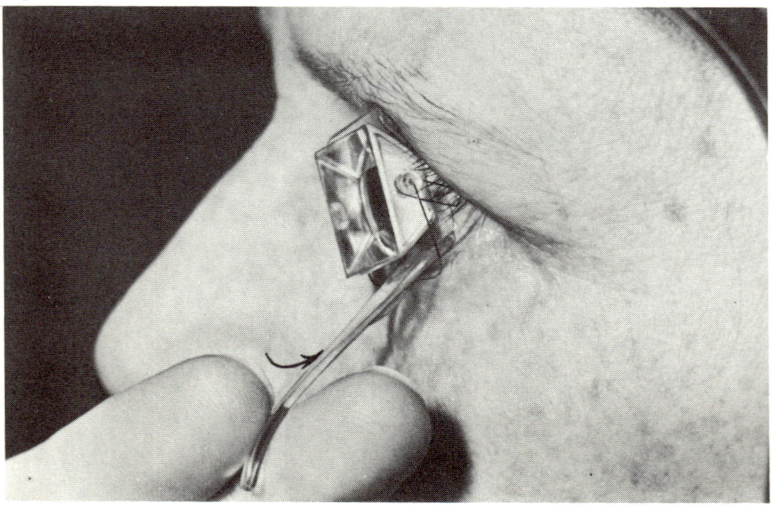

any of the types of primary glaucoma. It usually accompanies or follows disease or injury to the eye; an acute or chronic inflammation (iritis); a swollen hypermature cataract or dislocated lens; intraocular hemorrhage or tumor; or clots of blood blocking the veins or arteries of the eye. Although the symptoms and signs may be the same as those that occur with acute narrow-angle glaucoma, treatment is directed toward the causative condition.

CONGENITAL GLAUCOMA

Since a child's sclera is not completely developed, it still has an elastic element permitting the eye to stretch. When the eye becomes very large and looks like the eye of an ox or a cow, the condition is called buphthalmos, or hydrophthalmos. Congenital glaucoma, which may be discovered in the first three to four months of life, is due to faulty development of the outflow channels of the eye. One or both eyes may be large and bluish with a cloudy cornea (4.74).

ADULT GLAUCOMA

When pressure builds up in the adult eye as a result of blockage of the angle by the iris, structural drainage changes at the angle, or pigment, fluid may not be able to escape at the same rate that it enters. Unlike the child's eye, the adult eye will not stretch to any great degree. Any structure located between the internal pressure of the fluid and the sclera will lose its function and be destroyed. The structures in this location are the neurosensitive seeing parts of the eye, the retina

Fig 4.74

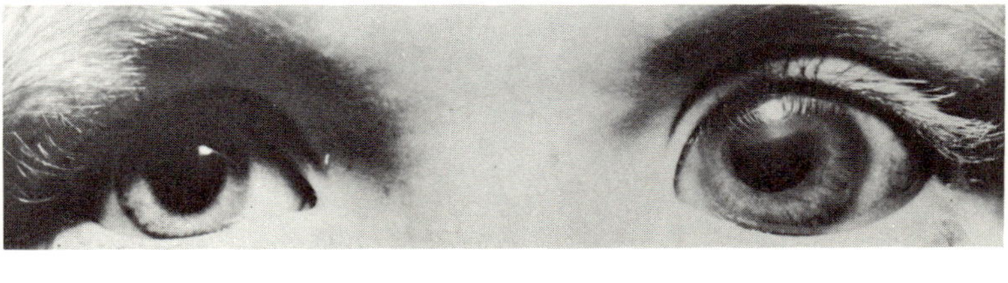

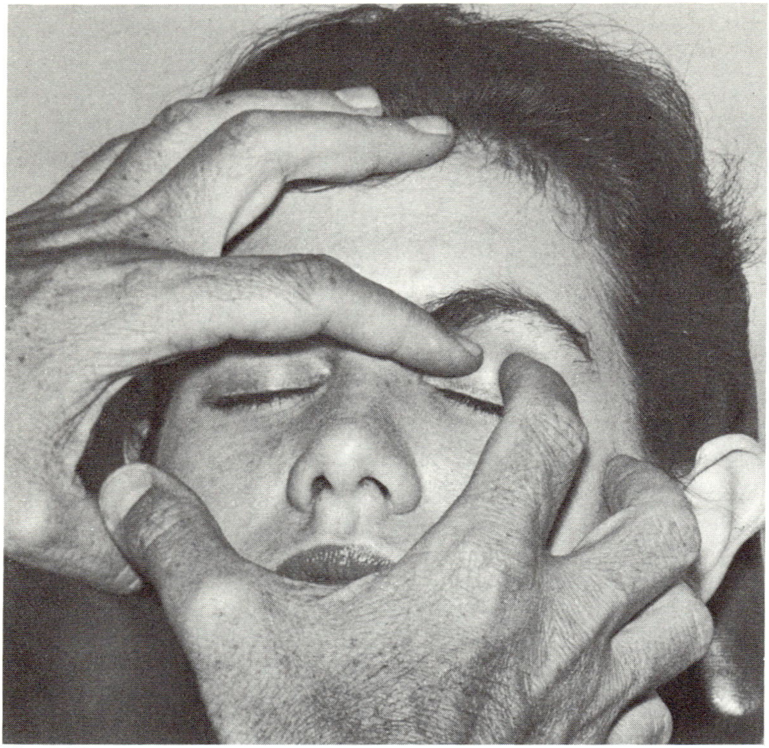

Fig 4.75

and optic nerve. Since they are made up of nervous system tissue, they never return to normal function if they have been destroyed. Vision lost by the effects of increased intraocular pressure never returns. If vision loss is total, the person is blind.

How Is Glaucoma Detected?

MEASURING PRESSURE OR TENSION IN THE EYE

Normally the eyeball feels like a soft rubber ball. Pressure in the eye measured by the fingers is called tactile (touch)

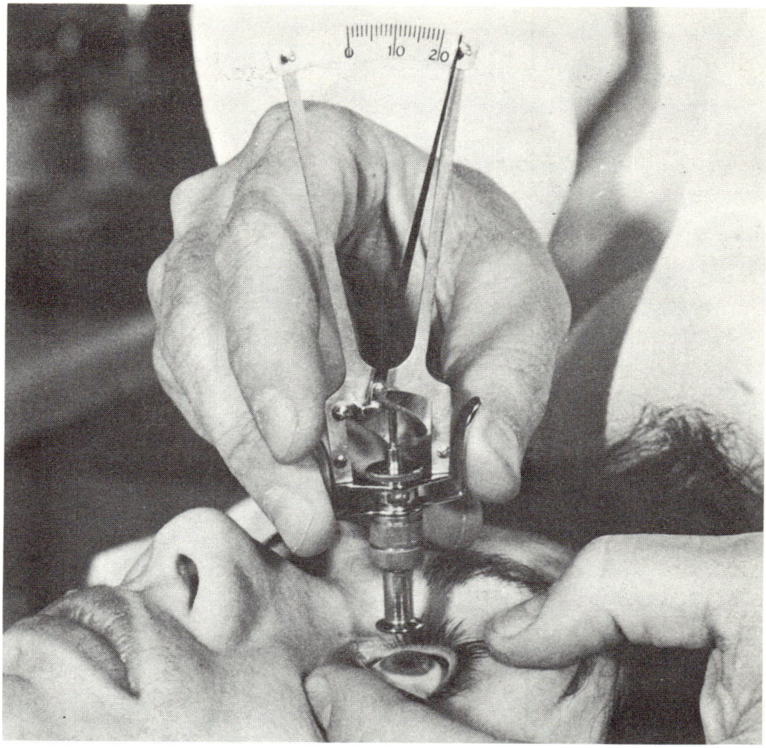

Fig 4.76

tension (4.75). This technique is like gauging the pressure in a tire by kicking it to see if it is hard or soft. For an accurate determination of pressure in a tire, however, an air gauge is used. Pressure in the eye can be determined accurately by a tonometer.

A tonometer measures intraocular tension. Before tension is recorded by the tonometer, drops are instilled to anesthetize the eye. These drops do not interfere with vision; they allow the tonometer to be applied to the cornea without causing pain or discomfort to the patient (4.76). The tonometer measures the amount of resistance of the eye to the indentation of a plunger. This resistance is then registered on a scale on the tonometer. By comparing this scale reading to a conversion scale table, the amount of pressure in the eye can be determined. Normal tension averages between 14 to 21 mm Hg. If pressure above 21 is recorded, the possibility of glaucoma should be investigated. Simplified tonometers, which do not accurately report the measurement in millimeters but only whether or not the tension is elevated, are sometimes used by general physicians.

The Schiotz tonometer is one measuring device, and the electronic tonometer is another. Still another, the applanation tonometer (4.77), can determine intraocular tension with the patient in an erect position. With this technique the cornea is not indented by a plunger, and the pressure is determined by direct scale reading. A hand applanation tonometer recently has been invented to take the pressure with the person in a reclining position. In a modernization of the old corneal staining technique, tension is determined by the imprint of the Maklakov tonometer applied to the cornea. Still more recently an instrument called a scleral tonometer has been devised; it is applied to the nonanesthetized eyeball and the pressure is recorded graphically.

PROVOKING A RISE IN PRESSURE IN THE EYE

If intraocular tension is slightly elevated during a routine eye examination, special provocative tests are utilized to determine the presence of glaucoma. Such tests are designed to create abnormal situations under which the function of the drainage mechanism can be studied. A patient, whose stomach is empty, may be given a quart of water to drink in five minutes; the tension in the eyes is then recorded for various time intervals and any rise in pressure is noted. Another test measures the rate of outflow of fluid from the eye by checking the intraocular pressure over a period of four minutes using the electronic tonometer. In yet another test the tension is measured with the pupils dilated and then constricted. On the basis of these tests plus visual field studies the ophthalmologist can determine if the patient has glaucoma.

GAUGING THE PROGRESS OF GLAUCOMA

The state of retinal function and the progress of the

Fig 4.77

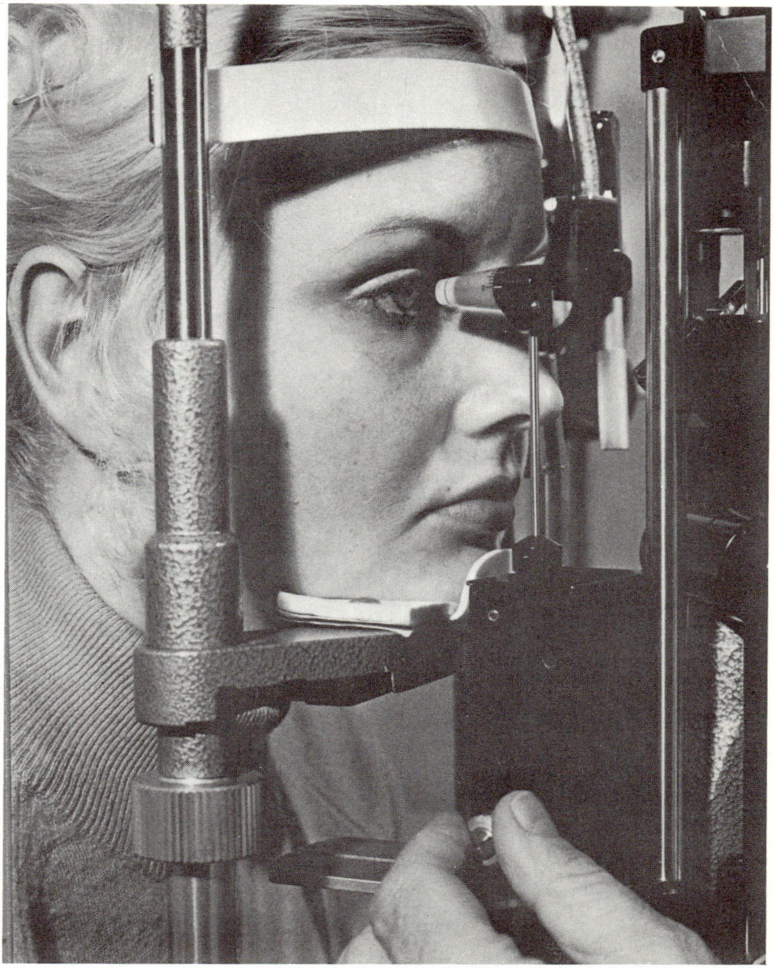

glaucoma are determined by testing visual acuity, by measuring the tension in the eye, by examining the fundus, and by studying the visual field.

Visual acuity. Although a diminution of visual acuity might indicate a loss of retinal function, central vision is often the very last part of sight to disappear in uncontrolled glaucoma.

Tension. By means of the tonometer intraocular tension is measured periodically. With controlled tension the retinal function usually remains normal, unless a low-tension glaucoma exists. With elevated tension retinal function may be impaired.

Fundus. With an ophthalmoscope the fundus is examined to see if pressure in the eye has affected the optic nerve, a structurally weak point in the eye. Glaucomatous cupping of the optic nerve head may impair sight.

Visual field. Studying the visual field is a fairly accurate way to determine retinal function. Central and peripheral visual field tests are carried out periodically in glaucoma patients or in those suspected of having glaucoma. Increased intraocular pressure can destroy the function of retinal nerve fibers, resulting in loss of vision or changes in the blind spot that is graphically revealed by visual field studies. This determination of sight loss, along with tension findings, often influences the course of therapy. One of the first signs of glaucoma is constriction of the peripheral vision. The patient may soon become aware that he cannot see cars or people passing beside him as he looks straight ahead; however, this same disability may be the chief complaint of a patient with advanced glaucoma. (For visual field tests, see pages 48–51.)

How Is Glaucoma Treated?

The medical and surgical treatment of glaucoma is aimed at decreasing the production of aqueous humor (shutting off the faucet), opening the angle (unplugging the drain), or producing a new filtration system (punching a hole in the tub) to allow fluid to escape from the eye and stabilize pressure. Thus sight is preserved in three ways:

- Slowing the flow of incoming fluid by tablets taken internally, by medication injected into the vein or instilled into the eye in the form of eye drops, or by an operation that destroys the ciliary body thereby inhibiting the production of aqueous humor.
- Enhancing the function of the drainage mechanism by eye drops called miotics that constrict or narrow the pupil, thereby opening the drainage angle of the eye and increasing the outflow of fluid. Since these drops also expose more of the iris structure to the aqueous humor and dilate the blood vessels, more aqueous humor leaves the eye.
- Creating a new filtration system by a painless operation that permits fluid to escape from the eye as fast as it forms.

Any or all of these three procedures may be performed to control increased intraocular tension and to preserve retinal function and sight. With newer medications only 20% or less of glaucoma patients require surgery in order to control tension.

Various glaucoma operations may be performed to filter fluid out of the eye by allowing aqueous humor to escape through some channel other than Schlemm's canal. The operation usually is performed under local anesthesia. An opening is made in the iris and sclera to permit fluid to leave the anterior chamber and to enter the subconjunctival space. The operation lasts about 30 minutes; only one eye is bandaged. The hospital stay is from two to four days. After that, eye drops and internal medication may be continued or terminated, depending on the state of the eye. The patient can plan on going back to work two or three weeks after surgery. Many of the hygienic measures suggested for cataract patients (page 136) are applicable to patients who have undergone an operation for glaucoma.

Misconceptions about Glaucoma

A high blood pressure and glaucoma go hand in hand. This is not true. A patient may have high blood pressure with or without glaucoma, or he may have both at the same time. However, drugs taken to lower blood pressure and to reduce body fluid may also be used to decrease intraocular pressure. Patients with high blood pressure and glaucoma should not have their blood pressure lowered rapidly. Glaucoma may progress more swiftly because of the decreased supply of blood to the optic nerves which results in visual field loss.

If your parents had glaucoma, you will have it. This may be true. Glaucoma has a dominant pattern of hereditary transmission. Parents of glaucoma patients usually have the disease, and children of patients could develop the disease. When near relatives have a history of primary glaucoma, children should be examined by their ophthalmologist. Persons past the age of 20, especially those who have a family history of glaucoma, should have their intraocular pressure measured during their yearly eye examination.

Every glaucoma patient goes blind. Glaucoma is *not* a blinding disease in 80% of all patients if it is diagnosed early and controlled adequately. No one should fear glaucoma today. He should feel fortunate that his condition has been discovered by his ophthalmologist during routine examination and that his sight can be preserved by proper therapy and by following these simple recommendations:*

1. Carefully follow your ophthalmologist's instructions and remember especially to return for reexamination at the appointed time.
2. Consult him at once if you see rainbow-colored halos around lights, if the eye becomes painful, or if vision is blurred or sight impaired in any way at all.

* As recommended in part by the National Society for the Prevention of Blindness, Inc.

3. Avoid excitement, anger, worry, fear, or disappointment as much as possible.

4. Take care that bowel movements are regular (every one to three days), and avoid straining.

5. Avoid tight-fitting collars, corsets, or belts.

6. If your occupation compels you to sit the entire day, take a long but not too tiring walk before and after work, or other forms of light nonstrenuous exercise.

7. Keep your teeth clean and healthy; pay attention to acute or chronic colds.

8. Limit drinking coffee and tea (not too strong) to one cup a day. Limit alcoholic drinks. Limit total daily fluid intake to six glasses a day, well spaced throughout the day.

9. Keep your bedroom well ventilated and at a moderate temperature (around 70°).

10. Avoid dark rooms as much as possible. Go to movies only if your ophthalmologist gives his permission. Remain at the movies for only one feature and, if possible, choose subjects that are not depressing or upsetting for you.

11. Do not use any drops or eye washes without consulting your ophthalmologist. They may be very harmful.

12. Be sure to let your family doctor know that you have glaucoma. Since many drugs that are commonly used for stomach disorders, skin diseases, and common colds, as well as numerous other conditions, can aggravate your glaucoma, other drugs can and should be prescribed by your physician.

13. Have a periodic (yearly) examination of your entire body by your family physician.

Glaucoma and Blindness

If a tire is overfilled with air, it may bulge in a weak spot and eventually have a blowout. If the eye is overdistended by glaucoma, it can and often does bulge in a weak spot. This area, which has a bluish-purple color due to the

Fig 4.78

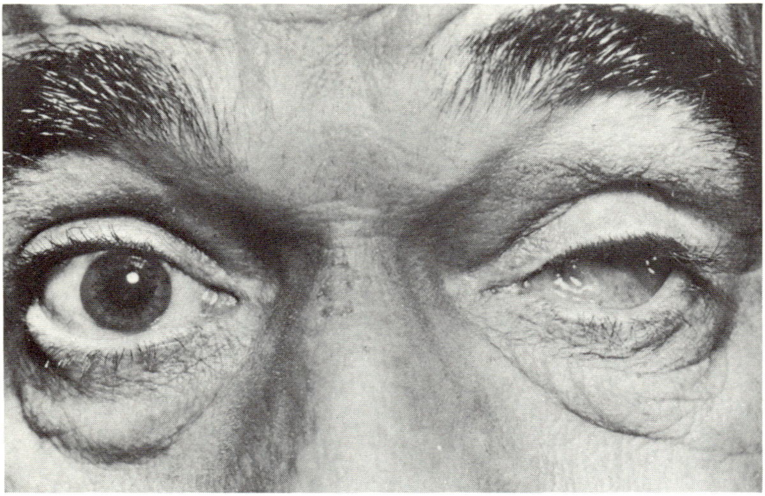

choroid being forced out of the weak spot, is called a staphyloma (4.57). Fortunately the eye rarely ruptures or has a blowout. The eye can also bulge backward at the optic nerve (glaucomatous cupping) or cause the nerve to pale and lose function with loss of sight. Total loss of sight due to glaucoma is called absolute glaucoma.

A chronically blind, painful eye resulting from glaucoma is irritable and often unsightly. The eye can be painlessly removed (4.78) under local or general anesthesia. The artificial eye inserted after enucleation is almost indistinguishable from the real eye (4.79).

The one-eyed person cannot overwork or strain his remaining eye no matter how much he uses it. The single eye will not get any stronger, but it will carry on the full job of sight efficiently. Although the one-eyed patient can manage well with vision in a single eye, he is handicapped by loss of side vision on the side of the lost eye and will not enjoy stereoscopic vision or depth perception. With two eyes, objects are viewed from different angles. These fields of vision

Fig 4.79

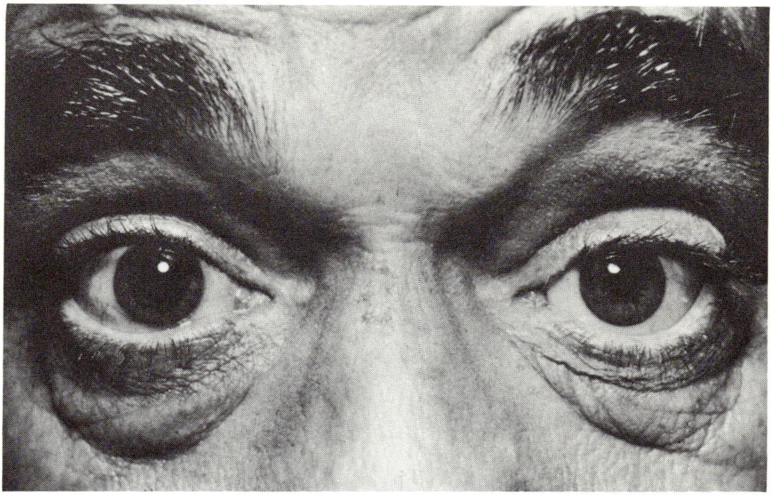

overlap or fuse to produce a single picture in the brain with depth perception. However, a one-eyed person can judge distance by parallax (the apparent change in the position of an object resulting from the change in the direction or position from which it is viewed), as well as by size, shape, and color differences. These same methods of judging distance are also used by the cross-eyed child who develops an amblyopic eye.

5

MYTHS AND MISCONCEPTIONS ABOUT THE EYES

IS THERE A BIT OF TRUTH in some of the common myths and misconceptions about the eyes, or are they modern "old wives' tales?" Will watching television, sunbathing, drinking, or smoking harm the eyes? Will eating carrots keep them healthy?

TELEVISION

Although black and white or color television itself is not harmful to the eyes, watching television often may exaggerate eye ache, fatigue, and headache in persons who need glasses. A viewer of television should sit at least 8 to 10 feet from the TV screen (a large screen is preferable) in a semilighted room with the light coming from behind him. The set should be at eye level and directly perpendicular, not at an angle, before the viewer to avoid distortion. The picture should be focused properly. When eye fatigue develops, viewing should stop. When children who sit directly in front of the set evidence refractive errors, they should be examined by their eye doctor. TV tubes, although a problem in the past, present no danger from x-ray or radiation injury.

LIGHTING

Fluorescent, Incandescent, Strobe, or Black Lights

Although these lights cause no definite eye disease, they may produce glare. Light rays reflecting from an object or shiny surface may result in symptoms of eyestrain, headache, or fatigue. This condition can be remedied by using indirect lighting, by adding a slight tint to the lenses of the glasses or buying tinted light bulbs, or by reducing light intensity.

Poor Lighting

Although poor lighting will not produce any organic eye disease, it may produce strain, fatigue, tiredness, or even headache.

Good Lighting

Light should come over the right shoulder for right-handed persons and over the left for south paws. Avoid poor reading habits. Do not read lying down in bed or in poor light. If reading oneself to sleep is a cherished habit, sit up and use proper lighting. Too much reading will not hurt the eyes, but it may cause eye fatigue.

SUN

Exposure of the eyes to bright sunlight may cause thermal dermatitis (heat burn or inflammation of the skin of the eyelids), thermal conjunctivitis (heat burn or inflammation of the conjunctiva), thermal keratitis (heat burn or inflammation of the cornea), or temporary or permanent sun blindness caused by burning destruction of the retina or macula.

Sun blindness has occurred in persons who have looked directly at an eclipse without proper eye protection. Those taking drugs such as LSD seem particularly prone to this type of burn. Smoked glass offers little protection to the eye if viewing an eclipse directly, but indirect viewing is safe. Make a pinhole in a piece of paper or cardboard. Hold this over another piece of paper or cardboard so the sun shines through the pinhole and produces the image of the eclipse on the second piece of paper.

Sunglasses or eye shields are recommended for sunbathers, skiers, or explorers on desert sands to help prevent excessive light from entering the eyes. Glasses of many different types and colors are available; they may be green calabar, green or gray ray ban, cosmo tan, or purplish to pink in color. By absorbing or reflecting unwanted rays, all tend to prevent the harmful rays of the sun from injuring the eyes. All colors are equally good, and the type of glasses chosen depends mainly on a person's color preference. Polaroid glasses also may be worn to reduce glare. Regardless of type, color, or cost, sunglasses cannot damage the eyes. However, poorly made lenses can cause eye fatigue. Sunglasses also may be worn to protect an irritated or inflamed eye or an eye with a dilated pupil. Since colored glasses decrease vision, they should not be worn when driving at night. This is the main reason why auto manufacturers do not use completely tinted windshields on cars.

Tinted prescription lenses are not as dark as sunglasses. A light tan to pinkish color is added to the lens with shades ranging from soft light No. 1 to 3. Photogray lenses, which darken in sun and bright light and lighten in reduced or dull light, have recently become available. These glasses should be used with caution when acute vision is required in rapid changes of lighting, for example, in entering and leaving tunnels. Unless one is extremely sensitive to light, tinted lenses are a needless expense. The shade or tinting of the lenses is not sufficiently dark to protect the eye from an excessive amount of light. And, in addition, the eye itself is normally well-equipped to screen out these excessive amounts of light.

SUNLAMPS

Sunlamps, like the sun, can produce thermal dermatitis, thermal conjunctivitis, and thermal keratitis. The same precautions should be observed. Colored goggles or eye occluders should be worn during exposure.

ALCOHOL

Drinking in moderation will not hurt the eyes. However, excessive drinking may cause redness of the eyes, double vision, or visual hallucinations such as pink elephants. Commercial brands of alcohol will not produce permanent eye disease; methyl (wood) alcohol produces blindness by its toxic destructive effect on the optic nerve.

SMOKING

Smoking in moderation will not affect the eyes, but the smoke itself may act as an irritant causing burning and redness. Excessive smoking—eight to ten cigars or three to six packs of cigarettes daily—may cause blindness.

EYEWASHES (COLLYRIA)

There are many commercial brands of eyewashes. None of them will cure an eye infection or an eye disease, but they will soothe an irritated eye and temporarily reduce redness. A persistently irritated or red eye should be examined by an ophthalmologist.

The eyecup is no longer used for treating or bathing the eye with eyewash. The cup may injure the eye while being inserted, may reinfect the eye by carrying infection from the lashes or margin of the lids into the eyeball itself, or may transmit the infection from one eye to the other, or from one person to another. Cotton or gauze compresses, soaked in either warm or cold water, may be used. They should then be discarded.

EYE MAKEUP

Eye makeup will not hurt the eyes unless some portion accidentally lodges in the eye. Some people may be allergic to chemicals in mascara, but the use of a nonallergic makeup often will remedy this situation. Although Vaseline® will not cause the eyelashes to grow or curl, it may be used to hold them in shape after curling.

RUBBING THE EYE

Rubbing the eyes usually will not produce blindness or an ocular defect. However, it can transmit an infection to the eye. In persons with diabetes or other diseases affecting the blood vessels, rubbing may rupture the fragile vessels, producing subconjunctival, retinal, or vitreous hemorrhages. If a foreign body is present in the eye, the sensitive cornea may be severely irritated or injured. Remember the old Chinese saying: "He who rubs his eyes with elbow won't get into eye trouble."

VITAMINS

Vitamin A

Vitamin A, which is contained in carrots, nourishes the retina by helping to produce visual purple. This agent is important in enabling one to see well at night. However, eating large quantities of carrots will not increase night vision unless a person is suffering from night blindness caused by a deficiency of vitamin A. Instead, a toxic reaction from too much vitamin A may result. The skin, palms of hands, and soles of feet may become orangish-yellow (hypercarotenemia or carotinoderma); the sclera or white of the eyes, however, remains white. Poor vision resulting from the need for glasses or from a cataract, glaucoma, or crossed eye cannot be prevented, retarded, or improved regardless of the quantity of carrots consumed.

Vitamin B

Vitamin B deficiency produces changes in the nerves leading to the cornea, eye muscles, and optic nerve. If these structures are inflamed or diseased, therapy with this vitamin may be indicated.

Vitamin C

Ascorbic acid deficiency may be responsible for hemorrhages in the conjunctiva or retina. Some physicians feel that Vitamin C in the form of lemon juice retards cataract development by improving lens metabolism; however, it will not cure or dissolve cataracts that have already developed.

Vitamin D

Lack of this vitamin was once thought to cause myopia or cataract formation. Regardless of the quantity of vitamin D taken internally, the course of myopia or cataract development will not be altered.

EYE EXERCISES

Eye exercises purportedly will cure myopia, hyperopia, or astigmatism and permit sight without glasses. This is not true. These refractive errors depend on the shape of the cornea and/or length of the eyeball; they can be corrected only by proper glasses that focus rays of light back to the retina, producing a clear picture. Exercises will not and cannot do this, neither can they improve cataracts, glaucoma, or color blindness.

Through the "exercises" advocated, the individual may concentrate more, may interpret images on test charts better, or may have the sensation of better vision. Subconsciously or consciously he is straining his eyes to focus on the object. Because of repetition, reeducation, and memory, he may think his sight has improved. Since this phenomenon is only temporary, it will not enable the patient to throw away his glasses. Persons without eye disease or refractive error who are nervous, fatigued, or slow readers may profit from this form of visual training. They may be taught to bring their vision up to par by more closely observing details, shape, color, contrast, and movements of objects.

The only real therapeutic value of eye exercises is in the

improvement of muscle imbalances, such as weakness of convergence, or the inability to turn the eyes toward the nose, as in the act of crossing the eye. In this instance, eye push-ups (the patient follows the head of a pin, pencil, or fingertip as it is brought toward his nose) tend to strengthen the medial rectus muscles and improve the near point of convergence, or his ability to cross his eyes. Other eye exercises called orthoptics help to improve binocular vision by stimulating depth perception.

EVIL EYE

We have all heard of the evil eye and its consequences. However, this is complete nonsense and purely an old witch's tale. There is absolutely no truth to the idea that an unborn child will develop birth defects if a pregnant woman is frightened by seeing snakes or crippled beggars while carrying the baby.

DEAD MAN'S EYE

There is an old wives' tale that whatever a person sees at the time of his death leaves a clear photographic imprint on his retina, which may be seen by looking closely into the dead man's eye. Another tale involves a dead rabbit. When the rabbit is killed, the last thing it sees also will leave an image on its eyes. This image may be seen if the rabbit's retina is developed like a film. Both of these tales are fabrications without any scientific foundation. Sight is a function of the brain. Chemical impulses are only temporarily recorded on the retina. From there they are transmitted to the brain for the actual interpretation of sight.

6

EMERGENCY EYE CARE

WHAT CAN AND CANNOT BE DONE IN AN ACUTE EYE EMERGENCY? With a badly injured eye it is better to do too little than too much. Use extreme care and gentleness in handling the lids and tissues. Do not use force to pry the lids apart. If both eyes are shut tightly, try opening the uninjured eye first. Remember a person who receives an eye injury is always upset and fearful of the outcome of this injury. Reassure him, but also warn him of the danger of squeezing his eyes. Remember, too, that the final outcome of his injury depends greatly on how and what first aid is rendered to the eye. When in doubt about either the nature of the injury or the kind of treatment required, apply a sterile eye pad or a clean handkerchief over the closed eye, and promptly take the injured person to an ophthalmologist.

FOREIGN BODIES

Cinders, pieces of steel, coal dust, eyelashes, wind, dirt, cigarette ashes, glass, or even an arrow may enter the eye (6.1). These bodies may lodge in the conjunctiva of the upper or lower lid, or on the cornea (4.23), and produce abrasions.

Symptoms

A scratchy sensation on blinking, pain, tears, light sensitivity, burning, or the comment that "something flew into my eye" may be the presenting complaint.

Treatment

Grasp the lashes of the upper lid with thumb and index finger. Lift the lid away from the eyeball, then gently pull it

Fig 6.1

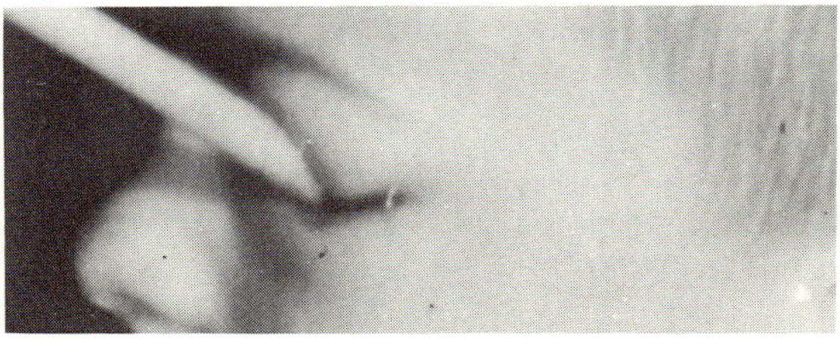

down over the lashes of the lower lid. The foreign body may dislodge or be trapped in the lashes of the lower lid. Tears formed during this procedure also help to wash out the foreign material.

Irrigate the eye with water to wash out the foreign body. This may be done by holding the eye open over a drinking fountain, by immersing the face and eye in a bucket or basin of water or under a shower, or by using an irrigating bottle (6.2).

Seventy percent of foreign bodies lodge on the conjunctiva under the upper lid. These bodies may be removed by turning back the lid. First have the person look down. Then grasp the lashes of the upper lid with the thumb and index finger of one hand. With the other hand insert a toothpick, match stick, or pencil on the skin surface of the upper lid (6.3). Then pull the lashes upward turning the lid over the toothpick and exposing the conjunctival surface of the upper lid (6.4). Remove the foreign body with a moistened sterile cotton applicator or the edge of a clean handkerchief. A for-

Fig 6.2

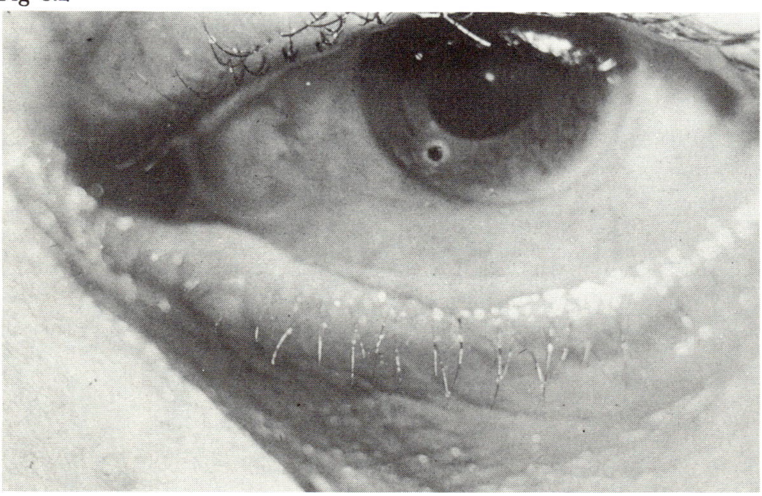

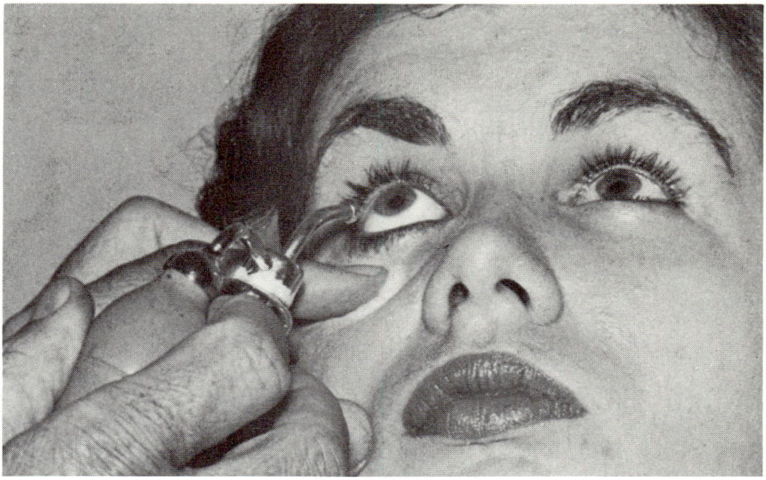

Fig 6.3

eign body lodging in the conjunctiva of the lower lid may be removed easily in a similar manner (6.5).

If these steps have not been successful in removing the foreign body, apply a sterile dressing over the closed eye,

Fig 6.4

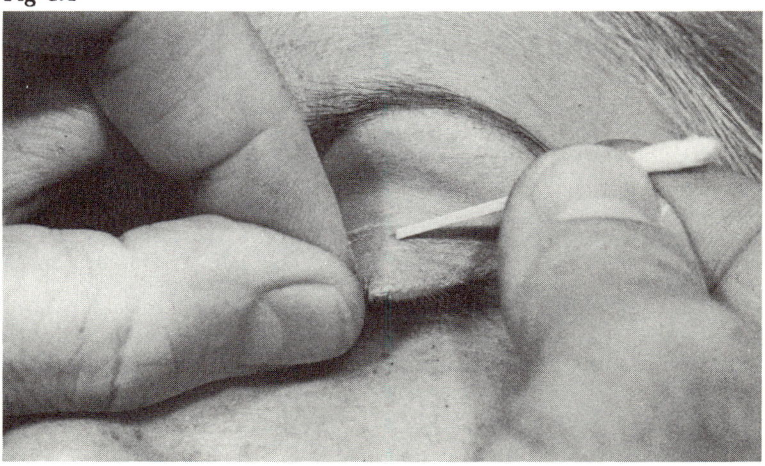

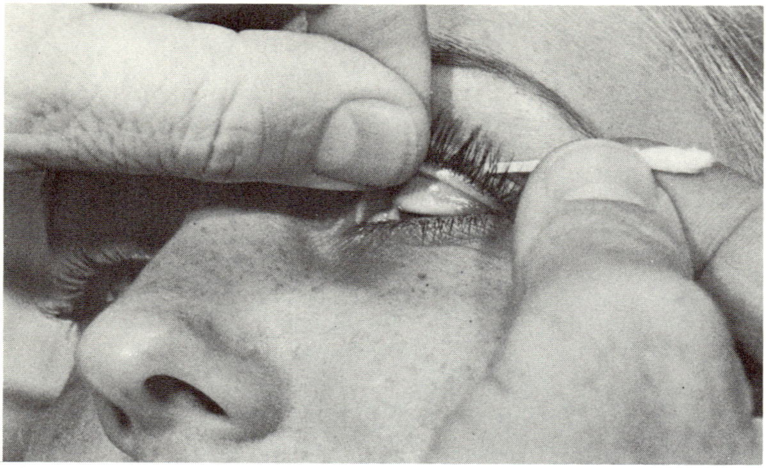

Fig 6.5

holding the pad in place with transparent tape or adhesive tape (6.6). If a sterile pad is unavailable, tie a clean handkerchief over the eye and head to prevent blinking and to relieve pain until medical care is obtained.

Do not rub the eye in an attempt to remove the foreign

Fig 6.6

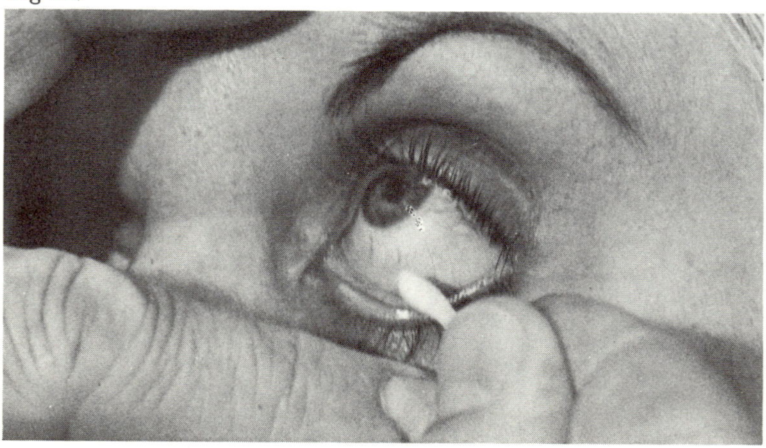

body. The foreign body may become embedded in the cornea (4.23). Do not use any eye medication "lying around the house" to provide relief unless the contents and prescribing instructions are actually known.

CHEMICALS OR DRUGS SPLASHED INTO THE EYE

Lye (4.25), gasoline, brake fluid, hair spray, blood, grease, or liquid medicine accidentally may be splashed or sprayed into the eye.

Symptoms

The patient complains of a painful, red, or burning eye that is sensitive to light and reports that he has splashed a chemical or medicinal agent in his eye.

Treatment

Without delay, copiously irrigate the eye for 10 to 15 minutes with warm or cold water (follow the instructions for irrigation listed under Foreign Bodies). Do not try to neutralize the chemical. *Do not rub the eye.* After bathing the eye, apply a sterile dressing or handkerchief, and then see a doctor immediately.

THERMAL (HEAT) BURNS

Exposure to the sun, a sunlamp, a welding arc, or other forms of ultraviolet light may cause thermal burns.

Symptoms

The person has been exposed to one of these agents four to eight hours earlier. His eyes and lids may be red or swollen; he may have painful burning eyes that are sensitive to light.

Treatment

Bathe the closed eyes with cold water compresses for five to ten minutes every two to three hours. Then apply a soothing ointment or lotion (Vaseline® or baby oil) to the skin and a dressing to both eyes to prevent blinking. An analgesic, such as aspirin, may be taken. If symptoms persist, see a doctor.

CONTUSIONS, BLOWS, OR INJURIES

A blow or injury to the eye may be delivered intentionally or unintentionally by a fist, hammer, baseball bat, or door.

Symptoms

The swollen and discolored eye or eyelids may appear red, black, or blue (6.7). There may be a subconjunctival hemorrhage (4.24), impairment or loss of vision, or severe pain.

Treatment

Apply cold water compresses of cotton, gauze, or cloth to the eye. Follow the instructions for applying cold water compresses listed under Thermal Burns. Cold compresses constrict blood vessels and tend to prevent further bleeding; they are more effective and much cheaper than beefsteak. Do not apply ice directly to the eye; it may produce further bleeding or injury. After 48 hours, alternately apply first warm and then cold compresses to the eye. By dilating and then constricting the blood vessels, this enhances the absorp-

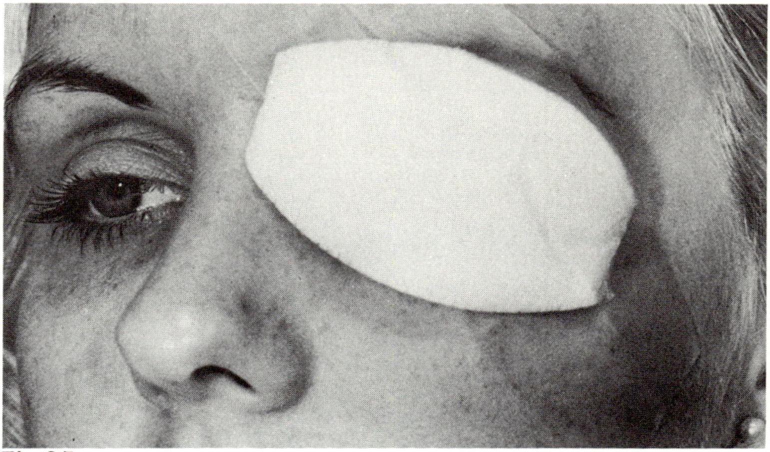

Fig 6.7

tion of blood. If loss of vision is present or pain persists, consult a doctor to rule out the possibility of a more serious underlying injury.

LACERATIONS (CUTS)

The eye or its surrounding structures may be cut by a razor, scissors, a knife, or other sharp implement.

Symptoms

Pain is usually present, and blood may be escaping from the cut.

Treatment

If the wound is located on the skin, cleanse it with soap and water (without getting the soap into the eye). Apply a sterile eye pad or clean handkerchief. Take care to exert no pressure with its application since the eyeball may have been cut. Consult a doctor immediately.

EXTERNAL INFECTIONS

Conjunctivitis, sties, and infected chalazions may be the result of external infections.

Symptoms

Itching, burning, watering, redness, swelling, or pain in or around the eye may be the presenting complaints.

Treatment

Bathe the eye with warm water compresses for five minutes at a time every two or three hours. Cold compresses applied for the same length of time often provide more relief. Do not use an eye cup; it encourages reinfection. Never patch the eye. Never rub the eye. Never use any eye ointments or drops "lying around the house"; they may be old and contaminated and completely inappropriate, or even harmful, to the present eye condition. Observe good hygienic habits. Be sure to use your own towel or washcloth, and make sure that no one else uses it. If your symptoms persist, see your doctor.

ALLERGIC REACTIONS

Symptoms

Itching, tearing, or swelling of the eyes after a bee sting, during asthma or hay fever season, or after the use of drugs are the chief complaints.

Treatment

Apply ice cold compresses to the closed eye. Follow the instructions listed under Thermal Burns. If there is no improvement, see a doctor.

SUDDEN PAINFUL RED EYE

Symptoms

A suddenly painful red eye that is sensitive to light, and loss of vision possibly accompanied by nausea or vomiting, might be caused by acute iritis or acute glaucoma.

Treatment

Bathe the closed eye with warm compresses. Follow the instructions listed under External Infections on page 172. Take an analgesic such as aspirin to relieve the pain, and then see an ophthalmologist immediately.

SUDDEN LOSS OF VISION

Symptoms

Sudden, partial, or complete loss of vision may be the result of a detached retina, intraocular hemorrhage, inflammation within the eye, spasm or clots in the central retinal artery or vein, or hysteria. The following case reports illustrate incidents of hysterical blindness:

> A soldier came into the hospital emergency room complaining of sudden loss of sight. When he was examined, his pupils were dilated and his eyes appeared to be normal. Upon further investigation, it was discovered that the soldier had called his home and learned that his young wife had given birth to a baby boy. Subconsciously, he was ap-

parently jealous of the baby and did not want to see him. Suddenly he became hysterically blind. With insight into his problem and certain stimulating medication, he made a sudden and complete recovery.

A man with an incurable bed-confining disease was separated from his wife and family. He became hysterically blind to obtain sympathy and attention from them. Again psychotherapy with insight into his problem resulted in his dramatic recovery.

Another person lost his sight hysterically, or psychologically, because he was in an automobile accident and wanted to gain financial compensation. His eyes were normal upon examination and he, too, recovered after becoming aware of his plight.

Treatment

Any pathologic eye condition should first be ruled out by a complete eye examination. After diagnosis of sudden visual loss has been established, proper ophthalmologic, medical, or psychiatric treatment can be given.

HEADACHES

Symptoms

A headache may indicate that a disease is present in or around the eye. If the eye is the cause, it may be due to improper correction of a refractive error, a low-grade iritis, glaucoma, muscle imbalance, or some other eye problem. Bad teeth or an infected sinus can refer pain to the eye.

Migraine headaches often run in families. They are one sided and preceded by visual, auditory, and systemic warning signs: flashes of light, zig-zagging lights, blurring of vision, or ringing in the ears followed by nausea and vomiting. High blood pressure, blood vessel spasm, or nervous tension may produce headaches or referred pain in and around the eye.

Treatment

A complete eye examination as well as a careful physical examination will reveal the cause of the headache, and proper therapy will be administered by the appropriate physician.

ORGANIZATIONS FOR EYE CARE

The National Society and state societies for the Prevention of Blindness, The American Association of Ophthalmology, The American Academy of Ophthalmology, The American Medical Association, and The Eye Bank Association of America all participate in programs to educate the public about eye problems and eye care. The American Foundation for the Blind is active in helping to restore blind or partially sighted persons to a useful life in their community. The Foundation may provide jobs, seeing-eye dogs, instruction in braille, or talking books. Talking books also may be obtained from local libraries or the Library of Congress in Washington, D.C. Special schools have been established for the visually handicapped. Sight-saving classes, where large printed books are used, are now being conducted.

Blindness is a disability. But today, thanks to the cooperation of the public and these organizations, the sightless person is neither alone nor helpless in a dark world.

Fig 6.8

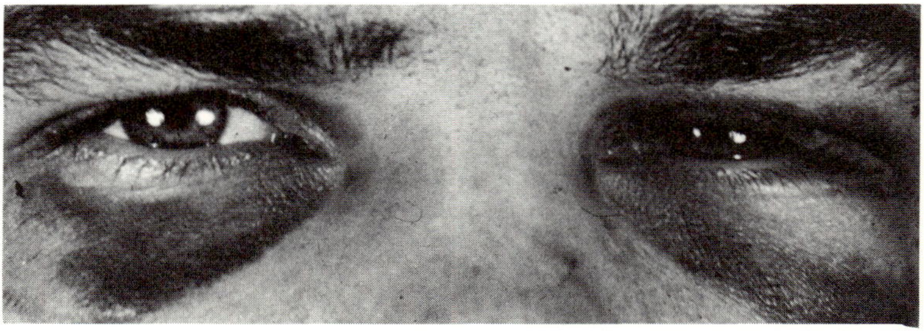

GLOSSARY

ACCOMMODATION: Adjustment of eye for seeing at different distances; accomplished by changing the shape of the crystalline lens through the action of the ciliary muscle.

ANTERIOR CHAMBER: Space in anterior portion of the eye, bounded in front by the cornea and behind by the iris; filled with aqueous humor.

AQUEOUS HUMOR: Clear watery fluid produced by ciliary body; fills chambers within front part of eye.

ASTIGMATISM: Defect of curvature of refractive surface of eye; rays of light from an observed object are not brought to a single focal point; images are distorted.

ATROPHY: Wasting away or diminution in size of a part either from lack of nourishment or a developmental abnormality.

BINOCULAR VISION: Coordinated use of two eyes to produce a single mental impression.

BLIND SPOT: Area inside eye not sensitive to light where optic nerve enters to supply nerve fibers to retina.

BULBAR CONJUNCTIVA: Part of conjunctiva covering anterior surface of eyeball.

CARUNCLE: Small fleshy body seen at inner angle of eye.

CATARACT: Opacity or cloudiness of lens, which obstructs passage of light rays and diminishes vision.

CHOROID: Vascular intermediate coat of the eyeball; supplies nourishment to other parts of eyeball.

CILIA: Eyelashes.

CILIARY BODY: Portion of vascular coat between iris and choroid consisting of muscles and blood vessels; focuses lens and manufactures aqueous humor.

COLOBOMA: A defect of the eye, especially a congenital fissure.

CONES: Together with rods, are receptors for optic nerve; they are the light-perceiving layer of the retina. Cones concentrated at macula are concerned with sharp vision and perception of shape.

CONJUNCTIVA: Mucous membrane lining eyelids and covering front part of eyeball except the cornea.

CONJUNCTIVITIS: Inflammation of conjunctiva.

CONCAVE LENS: Lens with power to diverge rays of light; also known as diverging, reducing, negative, myopic, or minus (—) lens.

CONVEX LENS: Lens with power to converge rays of light and bring them to a focus; also known as converging, magnifying, hyperopic, or plus (+) lens.

CORNEA: Anterior transparent portion of outer coat of eye through which light enters.

CRYSTALLINE LENS: Transparent, colorless nearly spherical body suspended in anterior portion of eyeball between aqueous and vitreous chambers; focuses rays of light.

CYCLOPLEGIC: Drug that paralyzes the power to focus and dilates the pupil.

DARK ADAPTATION: Power of the eye to adjust itself to a dim light.

DEPTH PERCEPTION: Ability to perceive the solidity of objects and their relative position in space (stereoscopic vision).

DIOPTER: Unit of measurement of strength or refractive power of lenses.

EMMETROPIA: Normal state of eyes in which rays of light reflected from a distant object are focused clearly on the retina when eye is at rest.

EPICANTHUS: Perpendicular fold of skin extending from base of nose to end of inner eyebrow.

EXOPHTHALMOS: Abnormal protrusion of eyeball.

EXTRINSIC MUSCLES: External muscles of eye, which move eyeball. Each eye has four recti and two oblique muscles.

EYE DOMINANCE: Tendency of one eye to assume major function of seeing, being assisted by less dominant eye.

FIELD OF VISION: Entire area that can be seen without shifting the gaze.

FOCUS: Point to which rays are converged after passing through lens; focal distance traveled by rays after refraction but before focus is reached.

FORNIX: Loose fold connecting palpebral and bulbar conjunctiva.

FOVEA CENTRALIS: Depression or pit in retina at posterior part of eye (temporal to the optic disk); most sensitive part of the retina.

FUNDUS: Interior back part of eye within its coats.

FUSION: Power of coordinating images received by each eye into a single mental picture.

GLAUCOMA: Disease caused by abnormally high intraocular tension within eye ("hardening of the eyeball").

HYPEROPIA: Farsightedness, a condition in which the theoretical focal point of rays of light reflected from a distant object lies behind the retina.

IRIS: Colored circular membrane suspended behind cornea and immediately in front of lens; regulates amount of light entering eye by changing size of pupil.

LACRIMAL GLAND: Gland lying in outer angle of orbit; secretes tears.

LACRIMAL PUNCTA: Small openings in lid margins near nose through which tears drain from eye.

LACRIMAL SAC: Dilated upper end of the naso-lacrimal duct.

LENS: Refractive medium having one or both surfaces curved; forms image by altering direction of light rays.

LIGHT ADAPTATION: Power of eye to adjust to variations in amount of light.

LIGHT PERCEPTION (LP): Ability to distinguish light from dark.

MACULA: Small specialized part of retina that receives detailed image focused upon it by crystalline lens.

MYOPIA: Nearsightedness, a condition in which rays of light reflected from distant objects are brought to focus before they reach the retina.

NEAR VISION: Ability to perceive objects distinctly at normal reading distance, usually considered to be approximately 14 inches from eye.

NIGHT BLINDNESS: Condition in which sight is good by day but deficient at night and in any faint light.

NYSTAGMUS: An involuntary rapid movement of the eye.

OCULIST or OPHTHALMOLOGIST: Licensed physician (M.D.) who specializes in diagnosis, refraction, and medical or surgical treatment of all conditions, defects, and diseases of eye.

OCULUS DEXTER (O.D.): Right eye.

OCULUS SINISTER (O.S.): Left eye.

OCULUS UTERQUE (O.U.): Both eyes.

OPHTHALMOSCOPE: An instrument with which the interior of the eye is examined.

OPTIC CHIASM: Crossing of fibers of optic nerves on ventral surface of brain.

OPTIC DISK: Head of optic nerve.

OPTICIAN: One who grinds lenses to prescription, makes up, and fits frames of glasses to the wearer.

OPTIC NERVE: Second cranial nerve; special nerve of sense of sight.

OPTOMETRIST: Licensed nonmedical practitioner who treats optical or muscular eye defects without the use of drugs or surgery. He may recommend glasses or eye exercises or the attention of ophthalmologist.

PAPILLEDEMA: Swelling of optic nerve at its exit from eye.

PERIMETER: Instrument for measuring field of vision.

PERIPHERAL VISION: Ability to perceive presence, motion, or color of objects outside of direct line of vision.

PINGUECULA: A benign fatty yellowish-gray elastic growth in the palpebral conjunctiva, near but not invading the cornea.

PHOTOPHOBIA: An intolerance to light.

POSTERIOR CHAMBER: Space between posterior surface of iris and anterior surface of lens; filled with aqueous humor.

PRESBYOPIA: A characteristic change of aging resulting in the inability of the eyes to focus on near objects.

PRISM: Piece of transparent material with two nonparallel plane surfaces used to displace an image.

PTERYGIUM: A growth consisting of vascular connective tissue of conjunctiva, which extends into cornea; occurs most often on the nasal side.

REFRACTION: (1) Bending of light rays as they pass from one transparent medium into another of different density. (2) Determining refractive errors of the eye.

RETINA: Innermost coat and perceptive structure of the eye.

RETINOSCOPE: Instrument for determining the refractive state of the eye by observing movements of lights and shadows across the pupil by light thrown onto retina from a moving mirror.

RODS: Together with cones, are receptors for optic nerve; the light-perceiving layer of the retina concerned with seeing light and motion.

SCLERA: Tough supporting outer coat of eyeballs; covers the posterior five sixths of the eye.

SLIT LAMP: Microscope providing a narrow beam of light like a searchlight; used with a microscope to examine anterior and posterior segment of eye.

SNELLEN CHART: Chart for testing central visual acuity in which letters or symbols are drawn to the Snellen scale of measurements; uppermost letter is designed to be read by the normal eye at 200 feet; rows of letters following that should be read at 100, 70, 50, 40, 30, 20, 15 and 10 feet.

SPHERICAL LENS: Segment of sphere refracting rays of light equally in all meridians.

STRABISMUS: Deviation of one of the eyes from its proper direction so that visual axes are not directed simultaneously toward the same object.

TARSUS: Framework of connective tissue that gives shape to eyelid.

TELESCOPIC GLASSES: Spectacles founded on principles of telescope; occasionally prescribed for improving very poor vision that cannot be helped by ordinary glasses.

TENSION, INTRAOCULAR: Pressure or tension of contents of the eyeball.

TONOMETER: Instrument for measuring intraocular tension.

UVEA (UVEAL TRACT): Entire vascular coat of eyeball; consists of iris, ciliary body, and choroid.

VISUAL ACUITY: Measurement of the ability of ocular system to distinguish detail.

VISUAL ACUITY: Measurement of the ability of ocular system to distinguish detail.

VISUAL PURPLE: Pigment of the outer segment of the visual rods.

VITREOUS: Transparent colorless mass of soft, gelatinous material filling four fifths of the eyeball behind lens.

VITREOUS FLOATERS: Fragments, such as threads, hairs, or spots, which float in the vitreous and can be seen because they interfere with light passing into the eye.

ZONULES: Attachments between ciliary muscle and equator of lens.